AA002335

MATERIALS RESEARCH SOCIETY
SYMPOSIUM PROCEEDINGS VOLUME 1067

Materials and Devices for "Beyond CMOS" Scaling

March 24-28, 2008
San Francisco, California, USA

Printed from e-media with permission by:

Curran Associates, Inc.
57 Morehouse Lane
Red Hook, NY 12571
www.proceedings.com

ISBN: 978-1-60560-845-7

Some format issues inherent in the e-media version may also appear in this print version.

CAMBRIDGE UNIVERSITY PRESS
Cambridge, New York, Melbourne, Madrid, Cape Town,
Singapore, São Paulo, Delhi, Tokyo, Mexico City

Cambridge University Press
32 Avenue of the Americas, New York, NY 10013-2473, USA

www.cambridge.org

Materials Research Society
506 Keystone Drive, Warrendale, PA 15086
http://www.mrs.org

©Materials Research Society 2008

This publication is in copyright. Subject to statutory exception
and to the provisions of relevant collective licensing agreements,
no reproduction of any part may take place without the written
permission of Cambridge University Press.

First published 2008

CODEN: MRSPDH

ISBN: 978-1-60560-845-7

Cambridge University Press has no responsibility for the persistence or
accuracy of URLs for external or third-part Internet Web sites referred to
in this publication and does not guarantee that any content on such Web sites
is, or will remain, accurate or appropriate.

Additional copies of this publication are available from:

Curran Associates, Inc.
57 Morehouse Lane
Red Hook, NY 12571 USA
Phone: 845-758-0400
Fax: 845-758-2634
Email: curran@proceedings.com
Web: www.proceedings.com

Materials and Devices for "Beyond CMOS" Scaling

Materials Research Society Symposium Proceedings
Volume 1067

San Francisco, California, USA
24-28 March 2008

TABLE OF CONTENTS

Theoretical Investigation of New Quantum-Cross-Structure Device as a Candidate Beyond CMOS 1
K. Kondo, H. Kaiju, A. Ishibashi

1T Memory Cell Based on PVDF-TrFE Field Effect Transistor 7
G.A. Salvatore, D. Bouvet, M.A. Ionescu, S. Riester, I. Stolichnov, R. Gysel, N. Setter

Atomistic Understanding of a Single Gated Dopant Atom in a MOSFET 12
G. Lansbergen, R. Rahman, C. Wellard, J. Caro, N. Collaert, S. Biesemans, G. Klimeck, L. Hollenberg, S. Rogge

Morphic Architectures: Atomic-Level Limits 18
R. Cavin, V. Zhirnov

Logic Devices with Spin Wave Buses - an Approach to Scalable Magneto-Electric Circuitry 33
A. Khitun, M. Bao, Y. Wu, J. Kim, A. Hong, A.P. Jacob, K. Galatsis, K.L. Wang

Author Index

Mater. Res. Soc. Symp. Proc. Vol. 1067 © 2008 Materials Research Society   1067-B03-01

Theoretical Investigation of New Quantum-Cross-Structure Device as a Candidate beyond CMOS

Kenji Kondo, Hideo Kaiju, and Akira Ishibashi
Laboratory of Quantum Electronics, Research Institute for Electronic Science, Hokkaido University, Sapporo, 060-0812, Japan

ABSTRACT

We propose a new quantum cross structure (QCS) device as a candidate beyond CMOS. The QCS consists of two metal nano-ribbons having *edge-to-edge* configuration like crossed fins. The QCS has potential application in both switching devices and high-density memories by sandwiching a few molecules and atoms. The QCS can also have electrodes with different dimensional electron systems because we can change the widths, the lengths, and the heights of two metal nano-ribbons, respectively. Changing the dimensions of electron systems in both electrodes, we have calculated the current-voltage characteristics depending on the coupling constants between a molecule and the electrode. We find that the conductance peak is much sharper in case of weak coupling regardless of dimensions of electron systems in electrodes, compared to strong coupling case. We also find that the conductance peak of QCS having electrodes with two-dimensional electron systems (2DES) is much sharper than that of QCS having electrodes with three-dimensional electron systems (3DES) in case of strong coupling because of quantum size effect of 2DES. These results imply that the QCS with the very sharp conductance peak can serve as the devices to switch on and off by very small voltage change.

INTRODUCTION

Nowadays, many researchers have paid attention to post-silicon devices [1,2]. One of the several post-silicon devices is a cross-bar memory device based on molecular devices fabricated by nanoimprint lithography, which has achieved the production of 30-nm half-pitch patterning [3,4]. However, today's production procedures such as nanoimprint lithography, and optical lithography, and electron-beam lithography, do not allow for the resolution to achieve sub-20-nm line-width structures.

Recently, we have proposed a double nano-"*baumkuchen*" (DNB) structure, composed of two thin slices of alternating metal/insulator nano-"*baumkuchen*" as a lithography-free nano-structure fabrication technology [5,6], and a quantum cross structure device [7,8]. The schematic illustration of the fabrication procedure is shown in figure 1. First, the metal/insulator (organic film) spiral heterostructure is fabricated using a vacuum evaporator including a film-rolled-up system. Then, two thin slices of the metal/insulator nano-"*baumkuchen*" are cut out from the metal/insulator spiral heterostructure. Finally, the two thin slices are attached together face to face so that each stripe is crossing. The DNB has potential application in switching devices or high-density memories, the cross point of which can be scaled down to ultimate feature sizes of a few nanometers due to the film thickness determined by the metal-deposition rate, ranging from 0.01 nm/s to 1 nm/s. We call one element of the DNB structure a QCS device that consists of two metal nano-ribbons having the edge-to-edge configuration shown in figure 1(c). The nano-ribbon electrodes can have different dimensional electron systems, such as 2DES or 3DES with their widths, lengths, and heights changed. We have already calculated the quasi-particle energy spectrum in two-dimensional (2D) electrodes using GW approximation [7]. Sandwiching an atom or a molecule between the two metal ribbons, the QCS device can detect the essential information of individual atoms or molecules. Also, we can make tunneling devices by sandwiching insulators, such as self-assembled monolayers (SAMs), DNA, or metal oxides. Therefore, we have reported about the calculation of electronic transport of tunneling in QCS devices [8]. In this paper, we have formulated the theory of electronic transport, which can deal with the difference of dimensionality of electrons in electrodes within the framework of the Anderson Hamiltonian, and we have studied the current-voltage characteristics of QCS devices with application of the theory. It is shown that the QCS devices are promising ones beyond CMOS quantitatively.

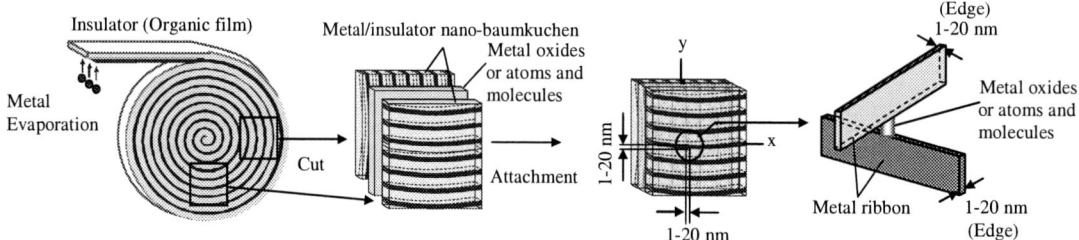

(a) Metal/insulator (organic film)   (b) Double nano-baumkuchen   (c) Quantum cross structure device
    spiral heterostructure by evaporation   by cut and attachment process

Figure 1. Schematic illustration of (a) a metal/insulator (organic film) spiral heterostructure, (b) a double nano-"*baumkuchen*" (DNB) structure, and (c) a quantum cross structure (QCS) device.

THEORY

We study the current-voltage characteristics of QCS devices with a molecule sandwiched between two metal electrodes. The molecule is assumed to have two energy levels.

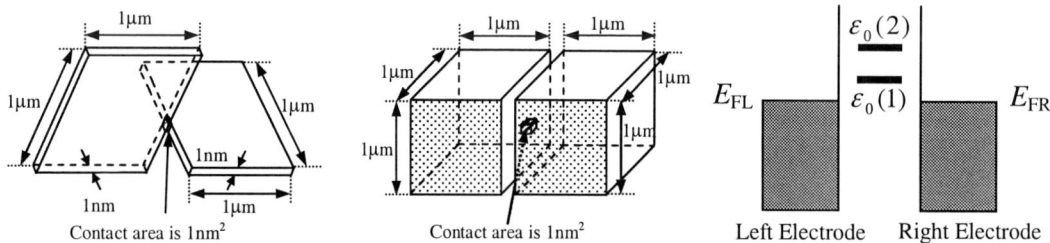

(a) QCS model with 2D electrodes   (b) QCS model with 3D electrodes   (c) Energy diagram for QCS devices

Figure 2. Schematic illustration of QCS device model (a) with 2D electrodes, and (b) with three-dimensional (3D) electrodes. The small shading box between the electrodes indicates a molecule position. (c) The energy diagram for QCS devices with 2D, 3D electrodes.

The models of QCS devices with 2D electrodes, 3D electrodes are shown in figure 2(a), figure 2(b), respectively. The energy diagram for the QCS device models is also shown in figure 2(c). We analyze the transport characteristics using the Anderson Hamiltonian, taking into consideration the dimensionality of electrons in both electrodes. In this case, the Anderson Hamiltonian is described as follows:

$$H = H_{\text{Electrodes}} + H_{\text{mole}} + H_{\text{t}},$$ (1)

where

$$H_{\text{Electrodes}} = \sum_{\alpha=L,R} \sum_{k,\sigma} \varepsilon_{k\sigma} \, c_{\alpha,k\sigma}^{+} c_{\alpha,k\sigma},$$

$$H_{\text{mole}} = \sum_{i,\sigma} \varepsilon_0(i) \, a_{i,\sigma}^{+} a_{i,\sigma},$$ (2)

$$H_{\text{t}} = \sum_{\alpha=L,R} \sum_{k,\sigma} \sum_{i,\sigma} (V_\alpha \, c_{\alpha,k\sigma}^{+} a_{i,\sigma} + h.c.).$$

$H_{\text{Electrodes}}$ is the Hamiltonian of both metal electrodes, $\varepsilon_{k\sigma} = \dfrac{\hbar^2 k^2}{2m}$, m is the free electron mass, and $\hbar$ is the Planck's constant h divided by 2π. The wave vector k is a 2D vector in 2D electrodes and a 3D vector in 3D electrodes. If you want to apply this theory to the interacting case, you need only replace $\varepsilon_{k\sigma}$ with quasi-particle energy spectrum obtained using GW approximation [7]. $c_{\alpha,k\sigma}^{+}$ and $c_{\alpha,k\sigma}$ are creation and annihilation operators for electrons of wave vector k and spin index σ in α electrode. α indicates the left or right electrode. The creation and annihilation operators obey the standard fermion anticommunication rules.

$$[c_{\alpha,k\sigma}, c_{\alpha',k'\sigma'}^{+}]_{+} = \delta_{k,k'} \delta_{\alpha,\alpha'} \delta_{\sigma,\sigma'}, \qquad [c_{\alpha,k\sigma}, c_{\alpha',k'\sigma'}]_{+} = 0.$$ (3)

H_{mole} is the Hamiltonian of a molecule sandwiched between both electrodes, and $\varepsilon_0(i)$ represents the i-th energy level of eigen-states of the molecule as shown in figure 2(c). We assumed the molecule had only two energy levels, $\varepsilon_0(1) = 0.5\,\text{eV}$, $\varepsilon_0(2) = 1\,\text{eV}$, estimated from Fermi levels E_{FL}, E_{FR} of each electrode. The Fermi levels E_{FL}, E_{FR} of each electrode were assumed to be equal and we used the value, 5.51eV, of the gold for the Fermi level. $a_{i,\sigma}^{+}$ and $a_{i,\sigma}$ are creation and annihilation operators for electrons of spin index σ in the i-th energy level. These operators also obey the standard fermion anticommunication rules like $c_{\alpha,k\sigma}^{+}, c_{\alpha,k\sigma}$.

H_{t} is the transfer Hamiltonian between the sandwiched molecule and each electrode, V_α is the transfer matrix between the molecule and α electrode. This value determines the coupling strength between each electrode and the molecule. Considering H_{t} as a perturbation, we have investigated the current-voltage characteristics from right to left electrode by changing these coupling strengths and the dimensions of electron systems in electrodes, using many-body perturbation technique. The electrons flow from left to right electrode. The transition probability $P_{L \to R}$ of electrons from left to right electrode is given by

$$P_{L \to R} = 2\sum_{k}\sum_{k} \frac{2\pi}{\hbar} |< Rk' | T | Lk >|^2 \, \delta(\varepsilon_{Rk'} - \varepsilon_{Lk}) f(\varepsilon_{Lk} - E_{FL})[1 - f(\varepsilon_{Rk'} - E_{FR})], \qquad (4)$$

where e is the elementary charge, $f(\varepsilon)$ is the Fermi-Dirac distribution function, $|Rk>, |Lk>$ are Bloch functions of electrons with eigen-energies ε_{Rk}, ε_{Lk} in right, left electrode, respectively, and T is so-called T-matrix, which is given in the following:

$$T = H_t + H_t G_0 T, \quad G_0 = \frac{1}{\varepsilon - H_0 + i\delta}, \quad H_0 = H_{\text{Electrodes}} + H_{\text{mole}}, \qquad (5)$$

where G_0 is the bare Green's function, and δ is the infinitesimal positive quantity. After tedious manipulation, the current I from right to left electrode is as follows:

$$I = e(P_{L \to R} - P_{R \to L})$$
$$= \frac{2e^2}{h} \int_{E_{FR}}^{E_{FR}+eV} d\varepsilon \sum_{i} \left(\frac{4\Gamma_L(\varepsilon)\Gamma_R(\varepsilon)}{(\varepsilon - \varepsilon_0(i))^2 + (\Gamma(\varepsilon))^2} \right) [f(\varepsilon - eV - E_{FR}) - f(\varepsilon - E_{FR})], \qquad (6)$$

where $\Gamma(\varepsilon)$ is the sum of $\Gamma_L(\varepsilon)$ and $\Gamma_R(\varepsilon)$, and $\Gamma_L(\varepsilon), \Gamma_R(\varepsilon)$ are coupling strengths between the molecule and left, right electrode, respectively, which are given in the following:

$$\Gamma_L(\varepsilon) = \pi D_L(\varepsilon)|V_L|^2, \qquad \Gamma_R(\varepsilon) = \pi D_R(\varepsilon)|V_R|^2, \qquad (7)$$

where $D_L(\varepsilon), D_R(\varepsilon)$ are densities of states of electrons in left, right electrode, respectively. These coupling strengths correspond to the broadening of the energy level of the sandwiched molecule induced by the interaction with itinerant electrons in each electrode.

RESULTS and DISCUSSION

We have calculated the transport characteristics at room temperature 300 K. Figure 3(a), 3(b) show the *I-V* characteristics of QCS devices with 2D electrodes, 3D electrodes, respectively, under the weak coupling condition. We regard the coupling as the weak coupling when the energy of coupling strength is smaller than that of the ambient temperature 27 meV. In this situation, the coupling constants V_L, V_R are 0.2 meV, corresponding to the coupling strengths $\Gamma_L(\varepsilon) = \Gamma_R(\varepsilon) = 1.57$ meV. We notice that both the results show the sharp steps at the positions of the energy level of the sandwiched molecule. Even at room temperature, these conductance peaks are sharp regardless of dimensionality of electrodes. The results are attributed to the discreteness of the energy level of the molecule. We can use a QCS device with 2D, or 3D electrodes as a switching device that works at very small voltage change under the weak coupling condition, which can become free from the so-called short channel effect of CMOS devices. Also, the flowing current of 0.5~1.0 μA is very small. Therefore, the power consumption is in the order of 10~100 nW at the device operation. The power consumption can be reduced until the much lower level by adjusting the coupling constants.

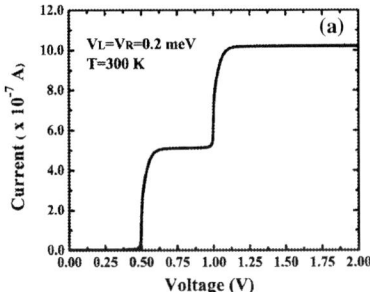 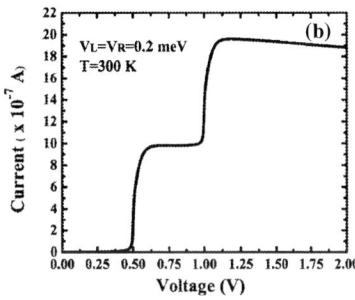

Figure 3. The current-voltage characteristics of QCS devices (a) with 2D electrodes, and (b) with 3D electrodes, under the weak coupling condition.

Figure 4(a), 4(b) show the *I-V* characteristics of QCS devices with 2D electrodes, 3D electrodes, respectively, under the strong coupling condition. We regard the coupling as the strong coupling when the energy of coupling strength is larger than that of the ambient temperature. In this situation, the coupling constants V_L, V_R are 1.0 meV, corresponding to the coupling strengths $\Gamma_L(\varepsilon) = \Gamma_R(\varepsilon) = 39.27$ meV. Compared to the weak coupling, the conductance peaks are smooth under the strong coupling condition for both QCS devices with 2D, 3D electrodes. However, the QCS device with 2D electrodes has much sharper peaks than one with 3D electrodes. The QCS device with 2D electrodes can still work as a switching device. This difference is attributed to 2D density of states changed due to the quantization of the energy of 2DES. As shown in figure 3,4, we find that the current of QCS device with 3D electrodes is larger than that of QCS device with 2D electrodes. This result is attributed to the larger density of states of 3D electrodes than that of 2D electrodes.

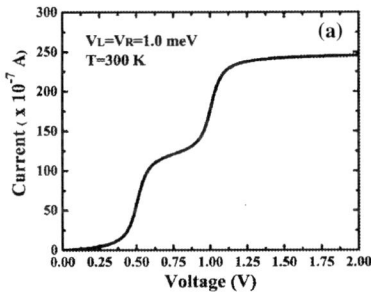 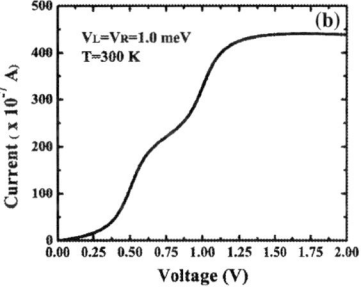

Figure 4. The current-voltage characteristics of QCS devices (a) with 2D electrodes, and (b) with 3D electrodes, under the strong coupling condition.

Finally, we consider the strong coupling limit. The strong coupling limit means that the energy of the coupling strengths $\Gamma_L(\varepsilon)$, $\Gamma_R(\varepsilon)$ are much larger than that of the ambient temperature. In this situation, we used the coupling constants V_L, V_R of 10.0 meV, corresponding to the coupling strengths $\Gamma_L(\varepsilon) = \Gamma_R(\varepsilon) = 3927$ meV. The energy level of the sandwiched molecule is almost continuous because of the wide broadening induced by the strong coupling with the itinerant electrons. Therefore, we expect the *I-V* characteristics like the ohmic contact without the molecule. The *I-V* characteristics of QCS devices under the strong coupling limit are shown in figure 5(a), (b). As expected, we can obtain the ohmic-like *I-V* characteristics, justifying the theory we have developed. To our knowledge, this is the first time to show explicitly that the *I-V* curve reduces to the ohmic-like one under the strong coupling limit within the framework of the Anderson Hamiltonian.

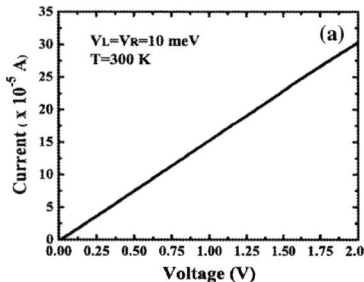 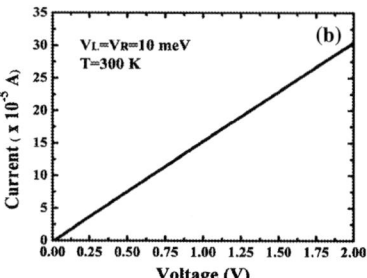

Figure 5. The current-voltage characteristics of QCS devices (a) with 2D electrodes, and (b) with 3D electrodes, under the strong coupling limit.

CONCLUSIONS

We have formulated the electronic transport theory, dealing with the difference of dimensionality of electrons in electrodes within the framework of the Anderson Hamiltonian. We have studied the *I-V* characteristics of QCS devices using the theory. We find that the QCS devices with 2D, 3D electrodes have sharp conductance peaks under the weak coupling condition even at room temperature. Therefore, we can expect the QCS device with 2D, or 3D electrodes to be the switching device that works at very small voltage change with the small current. The QCS device can operate at very small power consumption of 10~100 nW . These results imply that the QCS device is a promising one beyond CMOS. We consider that it is novel that this new QCS device is shown to be available for a switching device quantitatively, and that the theory is shown to easily apply to the interacting case.

ACKNOWLEDGMENTS

This research has been partially supported by Special Education and Research Expenses from Post-Silicon Materials and Devices Research Alliance.

REFERENCES

1. J. Chen, M. A. Reed, A. M. Rawlett, and J. M. Tour, Science **286**, 1550(1999).
2. Semiconductor Industry Association, *International Technology Roadmap for Semiconductors*, 2005 ed.
3. W. Wu, G-Y. Jung, D. L. Olynick, J. Straznicky, Z. Li, X. Li, D. A. A. Ohlberg, Y. Chen, S. –Y. Wang, J. A. Liddle, W. M. Tong, R. S. Williams, Appl. Phys. A **80**, 1173 (2005).
4. G. Y. Jung, W. Wu, S. Ganapathiappan, D. A. A. Ohlberg, M. Saifislam, X. Li, D. L. Olynick, H. Lee, Y. Chen, S. Y. Wang, W. M. Tong, R. S. Williams, Appl. Phys. A **81,** 1331 (2005).
5. A. Ishibashi, *Proc. Int. Symp. on Nano Science and Technology*, 44 (2004).
6. H. Kaiju, A. Ono, N. Kawaguchi, and A. Ishibashi, Jpn. J. Appl. Phys. **47**, 244 (2008).
7. K. Kondo and A. Ishibashi: Jpn. J. Appl. Phys. **45,** 9137 (2006).
8. H. Kaiju, K. Kondo, and A. Ishibashi: Mater. Res. Soc. Symp. Proc. **961,** O5.5.1 (2007).

Mater. Res. Soc. Symp. Proc. Vol. 1067 © 2008 Materials Research Society   1067-B03-02

1T Memory Cell Based on PVDF-TrFE Field Effect Transistor

Giovanni Antonio Salvatore[1], Didier Bouvet[1], Mihai Adrian Ionescu[1], Sebastian Riester[2], Igor Stolichnov[2], Roman Gysel[2], and Nava Setter[2]

[1]STI/IEL/LEG2, EPFL, EPFL STI IEL LEG2, ELB 237 (Bâtiment ELB), Station 11, Lausanne, 1015, Switzerland

[2]EPFL / STI / IMX / LC, EPFL, EPFL STI IMX LC, MXD 231 (Bâtiment MXD), Station 12, Lausanne, 1015, Switzerland

ABSTRACT

Interest in vinylidene fluoride (VDF) co-polymer with trifluorethylene (TrFE) P(VDF-TrFE) as ferroelectric material for memory application is driven by the prospect of having low cost and low operating voltage devices integrated on silicon and, at long term, migrate on flexible substrates. Some previous studies reported FET design using copolymers [1-8] but none of these structures were fully integrated on silicon wafers into a quasi-standard MOSFET fabrication process. We present for the first time the integration of a P(VDF-TrFE) (70%-30%) layer into a standard n-MOS transistor gate stack through a conventional semiconductor technology. This allows us to achieve a one-transistor (1T) Non Volatile Memory (NVM) cell. The operation voltage required for the 100nm organic ferroelectric thickness is less than 12V and a retention time ranging from few hours to few days is reported.

FABRICATION

A simple fabrication process is proposed in order to integrate the polymer and in order to make it as much as possible compatible with the CMOS technology (Fig.1). The substrate is a p-doped Si ($Na=10^{16} cm^{-3}$) with crystal orientation (100). We define the active area through Shallow Trench Isolation (STI). The source and drain regions are heavily doped ($Nd=10^{20} cm^{-3}$) by Phosphorous Oxychloride ($POCl_3$) and a thin film (10nm) of SiO_2 is thermally grown on the substrate to reduce the leakages and the screening in the ferroelectric gate. The P(VDF-TrFE) is prepared using a new recipe based on Methyl-Ethyl-Ketone, which helps to considerably reduce the film thickness and so the coercive field. The solution is spin coated and afterwards baked for 7 minutes at 137°C. The final layer thickness of 100nm has been measured and confirmed by an Atomic Force Microscopy (AFM). A thin gold layer (100nm) is sputtered to make the contacts and, after the last lithography, it is removed by wet etching. Based on this process, we have designed, fabricated and characterized transistors of different dimensions, ranging from 50μm down to 2μm (channel length and width) limited by the lithographic resolution available in our academic clean room.

Limitations of this fabrication process are the non-uniformity and the roughness of the polymer after the spinning process. The polymer has a roughness about 38nm with a peak value of 80nm and a root mean square of 7nm. This is a limitation in reducing the operating voltage that depends on the thickness of the ferroelectric layer. Future work will also focus on studying new deposition techniques that would guarantee much better uniformity and less roughness.

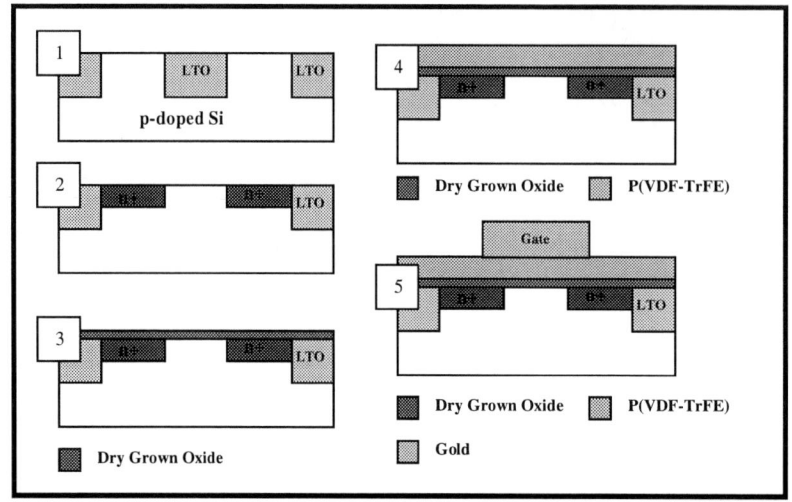

Fig.1 Simplified fabrication steps for P(VDF-TrFE) Fe-FET: <u>STI (Shallow Trench Isolation):</u> first lithography to define the active region, LTO deposition and CMP step to planarize the structure; 2) <u>Doping:</u> Drain and Source regions are doped by $POCl_3$; 3) <u>Gate oxide:</u> a thin layer (~10nm) of $SiO2$ is been grown; 4) <u>PVDF-TrFE deposition:</u> The polymer is spin coated and baked afterwards at 135°C for 5minutes; 5) <u>Gate electrode:</u> Gold is sputtered, patterned through UV lithography and then wet etched.

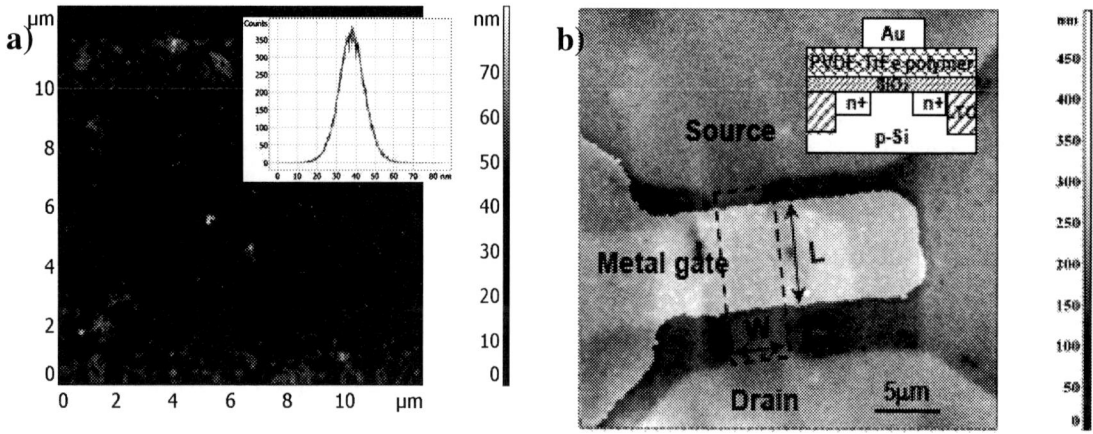

Fig.2 Atomic Force Microscope Images a) Polymer surface. The inset shows the roughness analysis: the average value is about 38nm with a root mean square about 7nm; b) SEM image of the Fe-FET. The image is tilted by 52°. Charging effect is clearly visible due to the dielectric P(VDF-TrFE) .

CHARACTERIZATION AND DISCUSSION

The ferroelectric properties of the polymer are preserved during and after the whole fabrication process. The polarization has been measured by sweeping-up the voltage on a P(VDF-TrFE) gate and on a $PVDF/SiO_2$ gate stack (Fig.3a-b). The difference in the slope of the two polarizations as function of the applied voltage clearly indicates the drop of voltage in the added oxide layer. The gate capacitance for SiO_2/P(VDF-TrFE) stack has been measured. P(VDF-TrFE) capacitance has been calculated from the total one considering the serie of the two layers.

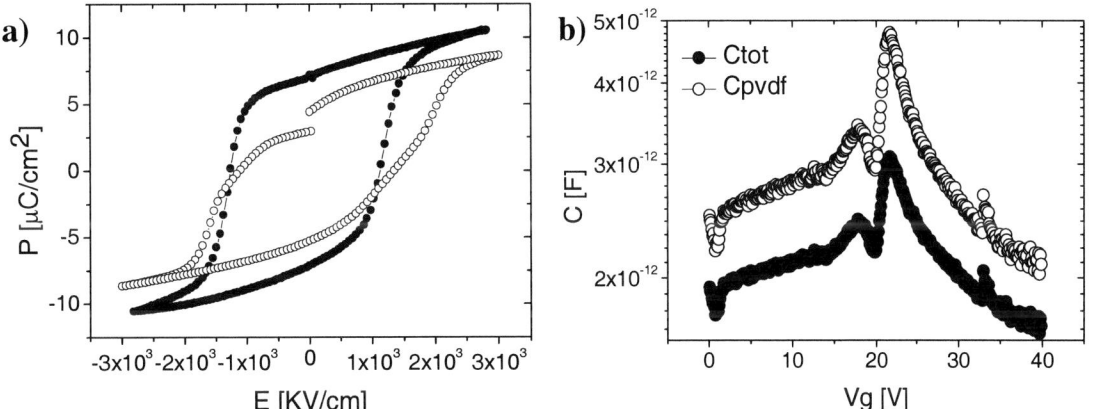

Fig.3 a) Gate Polarization loop. The plot using filled circular symbols is for pure PVDF layer while the plot using empty circular symbols corresponds to a PVDF/SiO_2 stack. For PVDF the remanent polarization is about 8μC/cm² and the coercive field, Ec, is about 1.3x10⁶V/cm. b) Capacitance measurement of a FET with L=W=40μm. The total capacitance is smaller than the one of PVDF layer because of the in-series gate oxide capacitance.

Experimental Id(Vg) and Id(Vd) curves for different channel lengths and widths show that our Fe-FET behaves similarly to a standard MOS transistor; the drain current being proportional to (W/L)Vd and a threshold voltage being visible (see Fig.4). The device shows very good hysteretic properties due to the polarization of the ferroelectric layer that induces a shift of the threshold voltage. This is translated in a controllable and reproducible memory behavior. The switching voltage is about 12-13V and this value is in fully agreement with the theory for which $V_{switch} = E_c * d$ (1), where E_c is the coercive field and d is the thickness of the ferroelectric layer (100nm in our experiments). For our polymer, we have measured an E_c=1.3*10⁶V/cm. This simple calculation does not take into account the drop of voltage onto the oxide layer. The saturation of the current is mainly due to the saturation of the polarization and so to the drop of the gate capacitance that drives the drain/source current. The change in current, due to the polarization of the ferroelectric layer, is in the order of 10⁵ and the off-state current is of the order of 10⁻⁹A (Fig. 4a). The hysteresis loop is fairly symmetric, which means that the read-out operation can be carried out for Vg=0V, value that is practical for low power application. Future work will be focused on reducing the operating voltage, hence in reducing the polymer thickness (according to (1)).

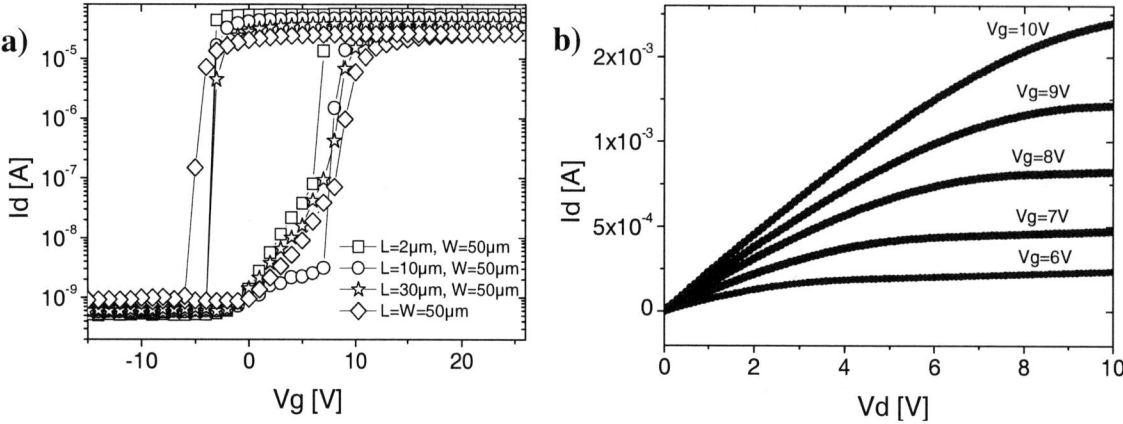

Fig.4 a) Id(Vg) experimental characteristics of a Fe-FET device with different channel lengths and W=50μm; b) Id(Vd) experimental characteristics of Fe-FET with L=W=50μm.

Ageing tests by accumulating multiple device cycling have also been performed. We cycled the Vg voltage (by ramping up and down its value, beyond +15V and −10V) and measured the drain/source current after successive series of cycles on the same device. After 10^3 cycles the current window is still large enough and the two states are clearly distinguishable. For the retention measurements we apply a voltage pulse on the gate and measured the time response of the drain/source current. It is found that the retention time depends on the FET dimensions. This can be as long as few days for the large transistors (L=W=50μm) and about 5 to 3 hours for smaller ones (for L=W=10μm and L=W=2μm, respectively). The mechanisms explaining the retention dependence on the device size are still under investigation; the role of defects size and density should be investigated for this purpose.

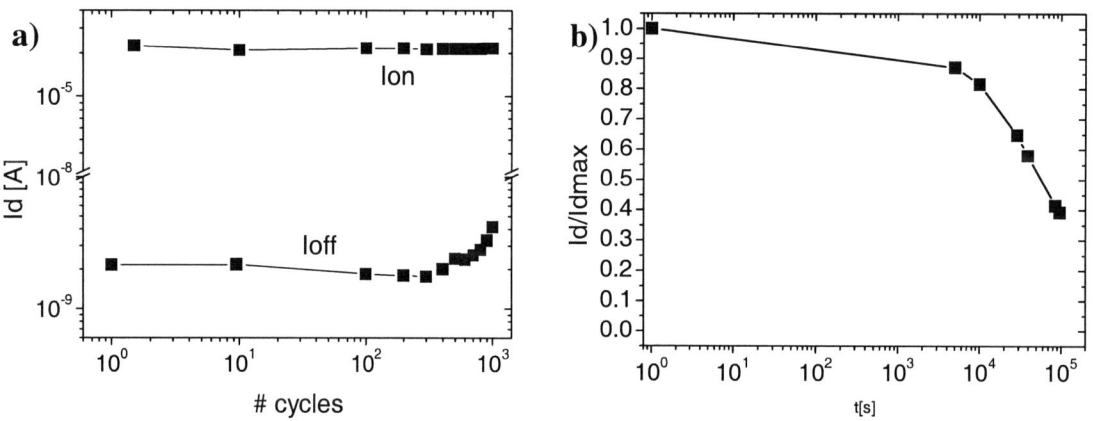

Fig.5 a) Typical result of ageing test on a Fe-FET device: after 10^3 cycles the two states are still well distinguishable; b) Retention measurement on the largest FET (50 x 50 μm²): after 10^5s (~1 day) the drain source current decreases by 60% of the initial value.

CONCLUSION

We designed, fabricated and characterized a new 1T Memory Cell based on a oxide /PVDF-TrFE gate stack field effect transistor. The fabricated device shows a memory window about 15V, good reliability and endurance up to 10^3 cycles. Retention time of the order of hours to a day has been demonstrated for our Fe-FET, which is promising for the use of such device as a memory in future electronic applications requiring data storage duration of similar order of magnitude (such as disposable electronic circuits).

Acknowledgements – The FP7 integrated project MINAMI and the CCMX Swiss research project are acknowledged for funding this work.

REFERENCES
[1] T. Furukawa et al., J. Appl. Phys., vol. 56, no. 5, Sep. 1984.
[2] S. H. Lim et al., J. Appl. Phys. v. 96, no.10 (2004).
[3] A. Gerber et al., J. Appl. Phys. v.100 024110 (2006)
[4] T. J. Reece et al., Appl. Phys. Lett. V. 82, no.1 (2003).
[5] R. C. G. Naber et al, Nature Material, v. 4 n.3 (2005).
[6] T. Sakoda, et al., Jpn. J. Appl. Phys., Part 1 40, 2911(2001).
[7] K. Kim et al., 2005 IEEE Symposium on VLSI Technology, (2005).
[8] S. Fujisaki and H Ishiwara, Applied Phys. Letts. 90, 162902(2007).

Mater. Res. Soc. Symp. Proc. Vol. 1067 © 2008 Materials Research Society

Atomistic Understanding of a Single Gated Dopant Atom in a MOSFET

Gabriel Lansbergen[1], Rajib Rahman[2], Cameron Wellard[3], Jaap Caro[1], Nadine Collaert[4], Serge Biesemans[4], Gerhard Klimeck[2,5], Lloyd Hollenberg[3], and Sven Rogge[1]

[1]Kavli institute of nanoscience, TU Delft, Delft, 2628 CJ, Netherlands

[2]Network for Computational Nanotechnology, Purdue University, West Lafayette, IN, 47907

[3]Center for Quantum Computer Technology, University of Melbourne, Melbourne, VIC 3010, Australia

[4]IMEC, Leuven, 3001, Belgium

[5]Jet Propulsion Laboratory, California Institute of Technology, Pasadena, CA, 91109

ABSTRACT

Current semiconductor devices have been scaled to such dimensions that we need take an atomistic approach to understand their characteristics. The atomistic nature of these devices provides us with a tool to study the physics of very small ensembles of dopants right up to the limit of a single atom. Control and understanding of a dopants wavefunction and its coupling to the environment in a nanostructure could proof a key ingredient for device technology beyond-CMOS. Here, we will discuss the eigenlevels and transport characteristics a single gated As donor. These donors are incorporated in the channel of a wrap-around gate transistors (FinFET). The measured level spectrum is shown to consist of levels associated with the donors Coulomb potential, levels associated with a triangular well at the gate interface and hybridized combinations of the two. The level spectrum of this system can be well described by a NEMO-3D model, which is based on a numerical tight-binding approximation.

INTRODUCTION

Isolated donors in silicon have received renewed attention in the last decade due to their potential use for quantum electronics [1-4]. An isolated donor forms a 3D Coulomb (thus truly atomistic) potential in the silicon lattice and can bind up to two electrons [5]. The isolated donors typically act as the binding sites for electrons with the information carried by either the electron-spin or -charge. The ability to perform operations with such structures is crucially provided by one (or more) gate electrodes around the donor site. Although much interest exists in the functionality that can be derived from isolated donors, experimental access to only a single donor has proven to be difficult [6-8].

Here, we will discuss resonant tunneling spectroscopy measurements on the eigenlevels of single As donors in a three terminal configuration, i.e. a gated donor which is a basic element for quantum electronics. These donors are incorporated in the channel of (p-type) prototype transistors called FinFETs. The local electric field due to the built-in voltage between the channel and the gate electrode forms a triangular potential at the interface. We will show that by means of spectroscopic measurements we can identify states to be associated with either the donors Coulomb potential, the triangular well or a hybridized combinations of the two. The theoretical framework used to describe this system is based on a tight binding approximation. The correspondence between the transport measurements, the theoretical model and the local environment of the donor provides a robust atomic understanding of actual gated donors.

FINFET DEVICE STRUCTURE

The FinFETs consist of crystalline silicon wires (fins) with large contacts patterned by 193 nm optical lithography and dry etching from Silicon-On-Insulator. After a boron channel implantation, a 100 nm polycrystalline silicon was deposited on top of a nitrided oxide (1.4 nm equivalent SiO_2 oxide thickness), then received a phosphorus (P) implant as predoping, and was patterned using an oxide hard mask to form a narrow gate. Next, we used high-angle arsenic (As) implantations as source or drain extensions, while the channel was protected by the gate and 50 nm wide nitride spacers and remains p type. Finally, As and P implants and a NiSi metallic silicide are used to complete the source or drain electrodes. The samples in this research have a gate length of 60 nm. Due to the relatively increased capacitance between the gate electrode and the corner regions of the nanowire, the later experiences a reduced potential. This so-called corner effect confines the source/drain-current to a narrow region at the very edges [9] which contains only a few As donor atoms see Fig. 1a. These donors originate most probably from transient enhanced diffusion at the Si/SiO_2 interface [10] out of the As source/drain contact extensions. In about one out of seven devices the distinctive resonances of the D^0 and D^- charge states of a single As donor can be observed in the transport measurements [7]. Note that we do not position the donors but rely on statistical chance for a donor to be present in the corner regions of the channel. The donors in the corner region are necessarily close to the gate interface.

a

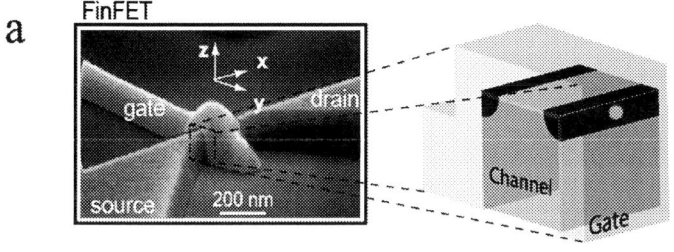

b

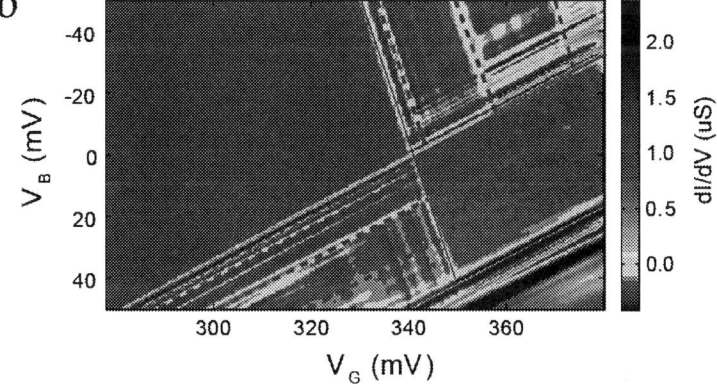

Figure 1. Device layout and stability diagram **a)** Colored Scanning Electron Micrograph of a FinFET device. Blow-up schematically shows channel/gate with current-carrying region (dark-blue) and donor atom (yellow dot). **b)** Stability diagram of a typical sample showing the D^0 charge state. The dashed red lines indicate the presence of excited levels.

LEVEL SPECTRUM OF A GATED DONOR

In this work, we will mainly focus on the level spectrum of the D^0 (single electron) charge state. These eigenlevels are determined from its measured stability diagram, i.e. a plot of the differential source/drain conductance (dI/dV) as a function of bias voltage (V_B) and gate voltage (V_G), see Fig. 1b. The D^0 state has its zero-bias peak at 340 mV gate voltage and is separated from the D^- state (just outside the graph) by the Coulomb diamond shaped region, where the current is Coulomb blocked. The source/drain conductance inside the conduction region of a charge state depends on the number of donor levels available for transport. The total electric transport actually increases as an excited eigenlevel enters the bias window defined by source/drain, giving the stability diagram its characteristic pattern [11] indicated by the dashed black lines. The red dots indicate the combinations of V_B and V_G where the ground state is at the Fermi energy of the drain and an excited state is at the Fermi energy of the source. It is the bias voltage V_B in this combination that is a direct measure for the eigenenergy of the excited state (e $V_{B,N} = E_N$), where E_N is the energy relative to the ground state and N is a label for the level). The excited states as determined in this fashion are depicted in Table 1. The eigenlevels do not match the levels of a bulk donor, but are heavily influenced by the electric field from the nearby gate electrode. The electric field is induced by the built-in voltage between gate and channel and can be estimated to be at around 21 MV/m. This estimation is based on a numerical solution of the Poisson equation and the charge distribution in a corner geometry. This is quite comparable to the Bohr field of the donor, ~ 30 MV/m.

Table 1. First three measured excited states of each sample versus the best fit to a tightbinding model (NEMO-3D). Also given are the donor depths (under the Si/SiO$_2$ interface) that were obtained from the measured charging energy versus the distance obtained from the TB-fit. The right-most column of the table lists the TB predictions for the local electric field and the standard deviation of the fit s. The experimental error per level across all devices is approximately 0.5 meV.

Device		E1 (meV)	E2 (meV)	E3 (meV)	Ec (meV)	d (nm)	F (MV m⁻¹)	s (meV)
10G16	*Exp.*	2	15	23	30	3.3		
	T.B.	2.2	15.6	23.0	-	3.3	37.3	0.59
11G14	*Exp.*	4.5	13.5	25	29	3.2		
	T.B.	4.5	13.5	25.0	-	3.5	31.6	0.04
13G14	*Exp.*	3.5	15.5	26.4	31	3.5		
	T.B.	3.6	15.7	26.3	-	3.2	35.4	0.17
HSJ18	*Exp.*	5	10	21.5	33	4.0		
	T.B.	4.5	9.9	21.8	-	4.1	26.1	0.63
GLG14	*Exp.*	1.3	10	13.2	35	4.7		
	T.B.	1.3	10	12.4	-	5.2	23.1	0.28
GLJ17	*Exp.*	2	7.7	15.5	33	4.0		
	T.B.	1.3	7.7	15.8	-	4.9	21.9	0.77

The eigenlevels of a gated As donor were calculated in an atomistic multi-million atom tight-binding approximation (NEMO 3-D) [12,13] as both a function of local electric field (F) and distance to the gate interface (d). The calculations include 1.4 million atoms corresponding to device volumes 30.4 x 30.4 x 30.4 nm. The corners of the FinFET are actually rounded with a radius of about 5 nm (about two times the Bohr radius of a bulk As donor), which justifies the planar nature of this model. Figure 2a shows the eigenenergies as a function of field for d = 4.3 nm as an example.

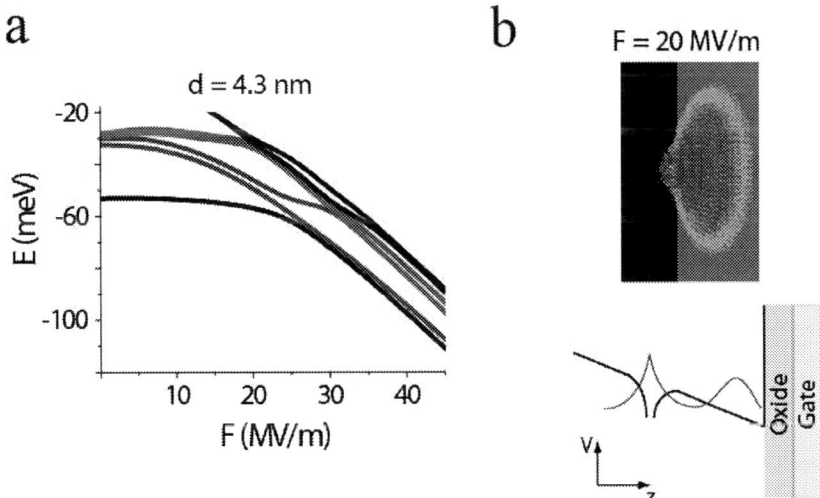

Figure 2. a) Eigenenergies (E) of an As donor 4.3 nm below a SiO$_2$ interface as a function of electric field (F) calculated with NEMO-3D. **b)** Wavefunction density of the ground state of an As donor at a donor depth d = 4.3 nm and local field F = 20 MV/m. The gray plane represents the SiO$_2$ interface. The ground state is a hybrid combination of donor-like and well-like states. Also indicated is a 1D scheme of the band diagram and a hybridized wavefunction.

Three electric field regimes can be distinguished. At the low field limit (F ~ 0 mV/m) we obtain the spectrum of a bulk As donor. In the high field limit (F~ 40 MV/m) the electron is pulled into the triangular well at the interface and the donor is ionized. In the cross-over regime (F ~ 20 MV/m) the electron is delocalized over the donor- and triangular well potential. Strong tunneling interaction between the two sites causes hybridization of levels characterized by the anti-crossing behavior of spectral lines. The ground state is a hybridized anti-bonding state of well-like and donor-like parts, see Fig. 2b.

The first three measured excited states of the D^0 state where fitted into the calculated spectrum with F and d as the two (independent) degrees of freedom. The six measured samples can be fitted quite well within the theoretical model, see Table 1. Taking into account the standard deviations of the fits and an estimated experimental error of 0.5\,meV we obtain a mean reduced chi^2 across the six samples of 0.92 (Note that chi^2 < 1 implies a good fit.)

LOCAL DONOR ENVIRONMENT

The local electric field (F) and donor depth (d) for each donor that follows from the tight-binding fit can be separately compared to independent determinations of their local environments. The charging energy of the D^- charge state is a direct measure of the donors distance to the gate interface (d). It follows from top of the Coulomb diamond between the D^0 and D^- as indicated in Fig 1b, as shown for all six samples in Table 1. The fact that the charging energy of the donors is reduced shows the donors are subjected to a (attractive) metallic screening. And, as can be readily observed, donors that are predicted to be closer to the interface by the TB-fit have a smaller charging energy. We explain the metallic behavior by the majority of Arsenic donors in the channel preferentially being segregated at the Si/SiO_2 interface [10], where it forms a (dipole) screening layer [14]. We can make a rough estimate of the reduction of the charging energy as a function of the donors distance to the interface by simply considering the donor as a small sphere which capacitance is reduced by the proximity of a metallic plate (the interface). This yields a surprisingly good result, see Table 1.

The local electric field consists of the electric field due to the built-in voltage and a contribution from the screening of the donor's dipole moment again by the gate interface. Figure 3 shows the positions of the measured donors in the F versus d plane as determined from the tight-binding fit. We find a trend for donors close to the interface to experience a higher local electric field, see Fig. 3, which can also be related to the afore-mentioned metallic-like screening at the Si/SiO_2 interface. The red curve shows a fit of the data-points assuming the donor nucleus and electron as point charges with a dipole arm a separating the two. This toy-model yields a very realistic dipole arm of a = 2.1 nm and captures the magnitude of the effect well, supporting our ideas on the metallic screening behavior of the interface.

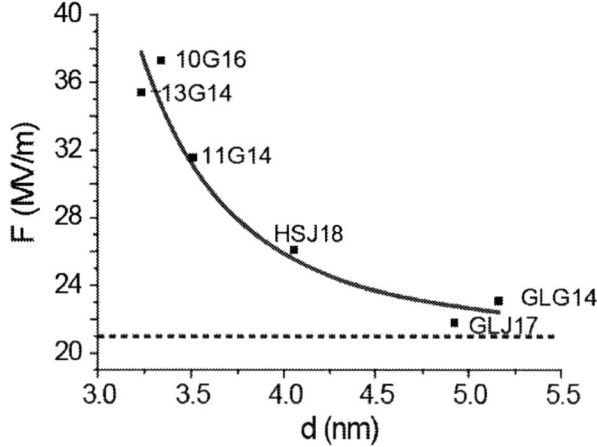

Figure 3. Local electric field F versus donor depth d as derived from the NEMO-3D model. The labels represent the corresponding devices. The F as expected from electrostatic modeling of the FinFET devices is indicated by the dashed line. The red curve is a fit of the data to a classical model of the interface screening as described in the text.

To make sure there is no significant effect of the aforementioned interface screening on the level spectrum, we modeled it at various relevant screening strengths and in refitted the data. We found only small changes in donor depth, in the range 0.1-0.4 nm, and local field changes less than a few percent.

CONCLUSIONS

We measured and explained the level spectrum of single gated As donors in a Silicon wrap-around gate transistor. The correspondence we find between the measured eigenlevels in the six samples and a multi-million tight-binding approximation shows we have a robust model for As donor states in a three-terminal geometry. The predictions that this model yields for the local field and donor depth correspond to the basic electrostatics of the nano-structured environment in which it is embedded.

REFERENCES

1. B.E. Kane, "A silicon-based nuclear spin quantum computer", Nature 393, 133 (1998).
2. R. Vrijen et al., "Electron-spin-resonance transistors for quantum computing in silicon-germanium heterostructures", Phys. Rev. A 62, 012306 (2000).
3. F. Ruess et al., "Realization of Atomically Controlled Dopant Devices in Silicon", Small 3, 563 (2007)
4. L.C..L. Hollenberg et al., "Charge-based quantum computing using single donors in semiconductors", Phys. Rev B 69, 113301 (2004)
5. This holds for shallow donors, see M. Taniguchi and S. Narita, "D- state in silicon", Solid State Commun. 20, 131 (1976)
6. L.E. Calvet, R.G. Wheeler and M.A. Reed, "Observation of the Linear Stark Effect in a Single Acceptor in Si", Phys. Rev. Lett. 98, 096805 (2007)
7. H. Sellier et al., "Transport Spectroscopy of a Single Dopant in a Gated Silicon Nanowire", Phys. Rev. Lett. 97, 206805 (2006)
8. S.E.S. Andresen et al., "Charge state control and relaxation in an atomically doped silicon device", Nano Lett. 7, 2000 (2007)
9. H. Sellier et al., "Subthreshold channels at the edges of nanoscale triple-gate silicon transistors", Appl. Phys. Lett. 90, 073502 (2007)
10. L.P. Kouwenhoven et al., in Mesoscopic Electron Transport, edited by L. L. Sohn, L. P. Kouwenhoven, and G. Schön (Kluwer, Dordrecht, 1997).
11. Z. Zhou et.al, "Dopant local bonding and electrical activity near Si(001)-oxide interfaces", J. Appl. Phys. 98, 076105 (2005)
12. G. Klimeck et. al., "Development of a Nanoelectronic 3-D (NEMO 3-D) Simulator for Multimillion Atom Simulations and Its Application to Alloyed Quantum Dots", Computer Modeling in Engineering and Science 3, 601-642 (2002).
13. G. Klimeck et. al., "Atomistic Simulation of Realistically Sized Nanodevices Using NEMO 3-D: Part I - Models and Benchmarks", IEEE Trans. Electron Dev 54, 2079-2089 (2007)
14. R. Kasnavi et al., "Characterization of arsenic dose loss at the Si/SiO2 interface", J. Appl. Phys. 87, 2255 (2000).

Mater. Res. Soc. Symp. Proc. Vol. 1067 © 2008 Materials Research Society
1067-B01-02

Morphic Architectures: Atomic-Level Limits

Ralph Cavin[1], and Victor Zhirnov[2]

[1]Research Operations, Semiconductor Research Corp., 1101 Slater Rd., Durham, NC, 27703
[2]Semiconductor Research Corp., 1101 Slater Rd., Durham, NC, 27703

ABSTRACT

In this paper, we consider a thought problem intended to force consideration of fundamental limits for energy sources, sensors, computing elements, and communication systems as fundamental system dimensions are reduced to the few micron regime. Design of integrated systems at this level are shown to literally require the allocation of atoms for the various functions. We argue that although there are no fabrication technologies for systems on this scale and the tradeoffs between system functions are extreme, systems on this scale might be feasible; given end-of ITRS technologies.

I. INTRODUCTION

A trend, synergistic with scaling, is the use of semiconductor technologies for diverse integrated systems applications. This trend is called 'Functional Diversification' and is characterized by the integration of non-CMOS devices such as sensors and actuators with traditional CMOS and other novel information processing devices. Functional diversification is empowered by continued dimensional scaling for integrated circuit devices and is fundamentally a cross-disciplinary activity. A specific application will drive specific technology requirements in areas such as system architectures, energy sources, sensors, packaging, etc.

We have imagined that it is desired to design a 'nanomorphic cell' whose function is, upon injection into the body, to interact with living cells, e.g. determine the state of the cell and to support certain "therapeutic" action. We stipulate that a microsystem should be on the order of the size of a living cell and have chosen a cube of ten microns on a side for the nanomorphic system. Our purpose is to examine the physical limits and trade-offs for each of the required system components, given severe volume limitations. In particular, the nanomorphic cell must have the capability to collect data on the living cell, it must analyze the data and make a decision on the type of cell; it must communicate with an external controlling agent; and possibly, take corrective action. Such a cell would need its own energy sources, sensors, computers, and communication devices, integrated into a complete system. In addition to the awesome challenges of designing and fabricating at the level of atoms, the nanomorphic cell need to be extremely energy efficient in its operations since only infinitesimal amounts of energy would be available to it. Fig. 1 below is a depiction of the nanomorphic cell.

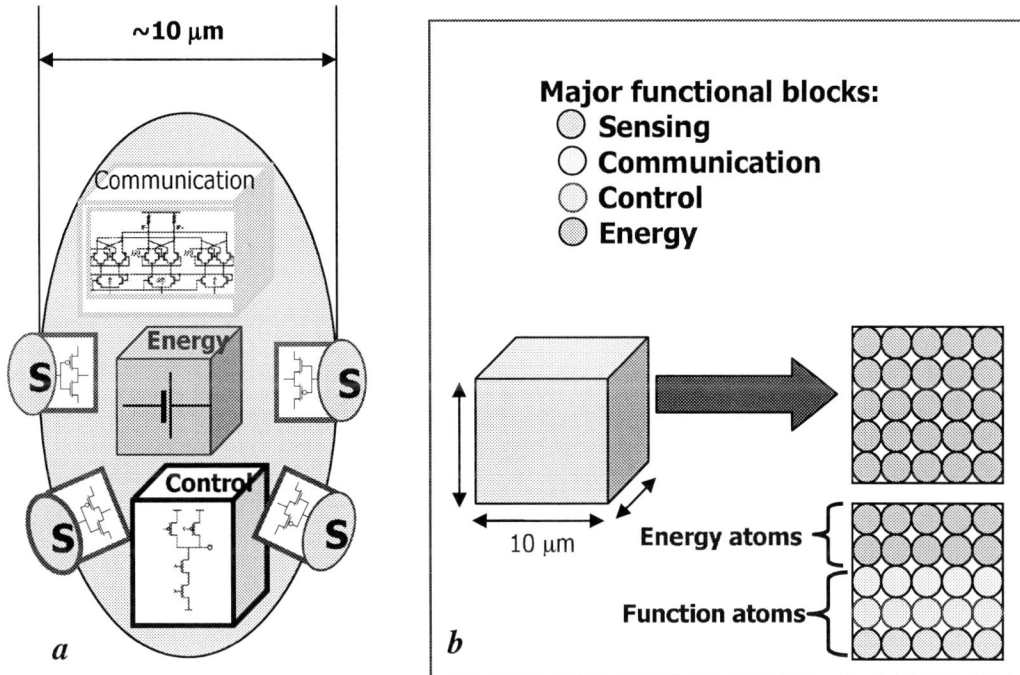

Figure 1. A Nanomorphic Cell: Generic architecture, constraints and trade-offs. Very limited space needs to by divided between sensors, power supply and electronic components (a). At this scale, every atom must play a role (b).

II. INTEGRATED MICRO-SCALE ELECTROCHEMICAL ENERGY SOURCES

In this section we investigate scaling limits and potential for micron-scale energy sources. There are a number of emerging applications where the space is the primary concern. Examples are on-chip integrated energy sources, integrated analytical microsystems, implantable diagnostics and drug delivery devices etc. Micro-scale energy sources are key enabler for extreme microsystems. In fact, the design options for such systems are constrained by limits imposed by energy sources.

II.1 Miniature galvanic cells

In a galvanic cell, the negative electrode loses metal atoms *M*, such as e.g. Li or Zn which are converted into ions, e.g. Li^+, Zn^{2+}, Al^{3+} that go into solution. The electrode becomes negatively charged, due to excess electrons that flow through the external connection. Thus, in the galvanic cell, '*atomic fuel*' is consumed to produce electricity: For every 1-2 electrons that flow through the external circuit, a metal atom must go into electrolyte solution as a Me^+ ion. When the supply of the metal fuel atoms is exhausted, the galvanic cell can no longer provide energy. Because the typical chemical bonding energy per electron is of the order of few eV, the typical potential difference V produced by such a system is ~ 1 Volt. Thus one atom of the atomic fuel produces ~eV of energy, and the total stored energy E_{stored} can be estimated as:

$$E_{stored} = eN_{el} \cdot N_{at} \cdot V \sim eN_{at}V \qquad (1)$$

e is the charge on an electron, N_{el} is the number of electrons released per atom, and N_{at} is the number of atoms in the metal electrode.

We can estimate the upper bound for the energy in an electrochemical source by using the fact that number of molecules (or atoms) in one mole of matter is given by the Avogardro's number N_A=6.02 x 10^{23} mol^{-1}, and that the atomic density in all solids, n_V, varies from 10^{22} to 10^{23} at/cm^3:

$$E_{max} \sim e \cdot N_A \cdot (1V) = 1.6 \cdot 10^{-19} \cdot 6.02 \cdot 10^{23} \sim 10^5 \frac{J}{mole} \qquad (2a),$$

or

$$E_{max} \sim e \cdot n_V \cdot (1V) = 1.6 \cdot 10^{-19} \cdot 10^{23} \sim 10^4 \frac{J}{cm^3} \qquad (2b),$$

and the *maximum energy stored in a 10-µm cube*:

$$E_{max} \sim e \cdot n_V \cdot (1V) \cdot l^3 = 1.6 \cdot 10^{-19} \cdot 6.02 \cdot 10^{23} \cdot (10^{-3})^3 \sim 10^{-5} J \qquad (2c)$$

Thus if the entire volume of the 10µm-size system is filled with the atomic fuel, a maximum of about 10^{-5} Joule would be available. This simple estimate is consistent with projections derived from practical scaling considerations.

The energy output is limited by the *number of atoms* available for conversion. When all atoms from the negative electrode are converted into ions, and the source is depleted, then continued use requires one of the following actions: (1) replace the cell; (2) recharge the cell by applying external electric energy and reversing the electrochemical reactions (converting ions back to atoms), or (3) "refill" the cell by replacing the electrode material (this option is realized in fuel cells). Of course in reality the 10 µm^3 volume must be shared with other components, and indeed a more accurate account for use of the volume would also include all essential components of the galvanic cell, i.e. anode, cathode, electrolyte and encapsulation.

Several comments can be made on the choice of 'fuel' materials for the extremely scaled galvanic cells:

1) Li appears to be one of best materials from the point of view of energy storage: it is has the largest standard electrode potential (-3.05 V) and it is the lightest solid material (ρ=0.53 g/cm^3). As result, Li-based batteries provide the highest gravimetric energy density (J/g), which is very important for e.g. lighter-weight portable applications and for transportation. However, it should be noted, that while Li-based batteries are primary choice for powering portable devices (because of their energy density), other kinds of batteries may better fit the constraints for extremely scaled bioelectronic systems. For applications, where space is the primarily constraint, other metals such as Al, Mg, Mn, and Ti outperform Li in volumetric energy density. Indeed for volumetric proficiency, Zn (the most popularly used primary battery at present) is comparable to Li.

2) The operating voltage is above 3 Volts in Li batteries, and this is attractive from the point of view of maximizing the total stored energy E_{stored}, which according to (1) is directly proportional to V. On the other hand, the energy dissipation in the load is proportional to V^2. For example, operation of electronic devices always involves charging and discharging of an equivalent capacitor C, and the corresponding energy dissipated in an elementary switching event is

$$E_{sw} = CV^2 \qquad (3)$$

The total number of the operations the system can perform with given energy supply (e.g. elementary binary switching events) is

$$N_{sw} = \frac{E_{stored}}{E_{sw}} \sim \frac{1}{V} \qquad (4)$$

Thus, the total number of switching events decreases as voltage increases. For these reasons the operating voltage should be below 1 V (for more on this see the following sections). If Li sources are used, they would require voltage conversion which consumes additional volume and dissipates additional energy.

3) Encapsulation may be the most important issue for the very small batteries, especially when electrode materials that are not compatible with water are used. Lithium violently reacts with water, and the need for a package (case) may impose a limit to practical miniaturization. To address this problem, caseless microbatteries have been proposed for bioimplantabe applications [1]. In fact, such caseless microbatteries consist only of two electrodes immersed in physiological fluids such as the subcutaneous interstitial fluid, blood, serum etc.

II. 2. Miniature Bio Fuel Cells

The big attraction of fuel cells is that their energy capacity is not limited as long as energy (in chemical form) is supplied from outside. For semiconductor bioelectronics applications, it is attractive to use a small amount of energy stored in a biological organism to power the nanomorphic microsystem. One important 'biofuel' is glucose. The glucose-O_2 biofuel cell utilizing carbon fiber electrodes is a promising candidate for micropower source [1, 2]. There is a potential for CNT as electrode material in extremely scaled bio-fuel cells.

Another important "molecular fuel" in biosystems is adenosine 5'-triphosphate (ATP) - $C_{10}H_{16}N_5O_{13}P_3$. The ATP energy storage density is ~6.6 x 10^4 J/kg or ~3 x 10^4 J/mole [3]. This corresponds to ~0.3 eV per ATP molecule, which is comparable to ~ 1 eV in the galvanic cell. However, in bio-systems, the ATP is dissolved to a typical concentration of 1-10 mM [3], and thus the maximum stored energy is reduced to ~0.3-3 J/cm^3.

II.3. Miniature supercapacitors

The micro-scale electrochemical sources described in the previous sections have power output in μW range, which could be sufficient for some applications (e.g. sensing, logic, control), but too small for others, e.g. signal transmission. Integration of a micro-scale battery with micro-scale supercapacitor may help to boost the power output of the system.

Supercapacitors are electrochemical energy storage devices, which are close relatives' to the galvanic cell. There are two energy storage mechanisms in supercapacitors. The first mechanism results from the formation of an electrical double layer at the electrode/electrolyte interface. The second energy storage mechanism is due to the voltage dependent fast electrochemical (faradaic) reactions occurring at the electrode surface between electrode atoms and ions of the electrolyte.

The capacitance of a double-layer supercapacitor can be estimated from a standard equation:

$$C = \frac{\varepsilon_0 \varepsilon A}{d} \qquad (5)$$

where ε_0=8.85x10^{-14} F/cm, ε is the relative dielectric constant of the interpolate layer, d is the thickness of the interpolate layer, and A is the surface area of the capacitor plates.

The basic concept of supercapacitors is to maximize C by creating the minimum possible d in (5) due to the use of the electrical double layer at the metal-electrolyte interface. For concentrated aqueous electrolytes, d=0.5-1 nm [4]. By using this value and the dielectric constant inside the double layer, ε~10 for aqueous electrolytes [4], (5) gives specific capacitance (the capacitance per unit area, C_s=C/A) as C_s~10-20 μF/cm^2.

The faradaic capacitance (pseudocapacitance) is defined as $C_f = \dfrac{q_f}{V}$, where q_f is the charge transferred during electrochemical reaction. The maximum value of q_f is limited by the surface density of atoms n_s (which in all solids varies from ~10^{14} at/cm^2 to ~10^{15} at/cm^2):

$$q_{f_{max}} \sim e \cdot n_s \sim 1.6 \cdot 10^{-19} \cdot 10^{15} \sim 10^{-4} \frac{C}{cm^2} \qquad (6)$$

For V~1 V, the faradaic capacitance per unit area C_f~100 μF/cm^2.

The total capacitance can be further increased by increasing surface area using porous electrode materials with high specific surface a_s (total area per unit mass). Different forms of activated carbon offer the highest specific surface – on the order of 1000 m^2/g. The upper bound for the specific surface of carbon can be estimated to be 2600 m^2/g based on atomic packing considerations of graphene sheets in graphite [5].

The energy, stored in the capacitor is

$$E = \frac{CV^2}{2} \qquad (7)$$

The maximum voltage V_{max} of the electrochemical double layer capacitors is limited by the electrolysis threshold voltage. For aqueous electrolytes, V_{max}=1.2 V, for organic electrolytes V_{max}~2.5 V, and for ionic liquids – organic molten salts V_{max}~4 V.

The power delivery by discharge of a capacitor is:

$$P = \frac{E}{t} \qquad (8),$$

where t is the energy release time, which depends on the time constants of electrode reactions (e.g. formation/relaxation of double barrier or redox reactions) t_{el}, and the capacitor discharge time t_C:

$$t = t_{el} + t_C \qquad (9)$$

The electrode reaction time is t_{el}~10^{-8} s for double-barrier formation process, and t_{el}~10^{-4}s for faster redox reactions [22]. The capacitor discharge time t_C~$2RC$ (double-layer capacitor, roughly 90% V-discharge, 99% E-discharge), where R is the resistance of the electrolyte, which can be calculated using conventional formula:

$$R = \frac{l}{\sigma A} \qquad (10)$$

In (13) σ is the conductivity of electrolyte, $A=L^2$ is the planar electrode area, and l is the electrolyte layer thickness. Electrolyte layer thickness must be at least larger than $2x$ the double layer thickness, and therefore following [5] assume l=2 nm for the hypothetical best case scenario. For L=10 μm and σ~0.7 S/cm (concentrated aqueous electrolyte), the resulting resistance R~0.3 Ω. For the ionic liquids s~0.01 S/cm, which results in R~20 Ω.

Characteristic parameters driving the performance of supercapacitors are given in Table I, along with estimates using (7)-(10) for the energy and power delivery by a 10 μm-sized supercapacitor. Note that the upper bound for power is ~1 W.

Table I. Estimated parameters of near-ideal 10 µm-sized double-layer and faradaic supercapacitors

Capacitor	**Double-layer**		**Faradaic**	
Electrolyte	Aqueous	Ionic	Aqueous	Ionic
V_{max} *(Volt)*	~1.2	~4	~1.2	~4
C_{max} *(F)*	~$6 \cdot 10^{-7}$	~$6 \cdot 10^{-7}$	~10^{-5}	~10^{-5}
R_{min} *(Ω)*	~0.3	~20	~0.3	~20
t_{el} *(s)*	~10^{-8}	~10^{-8}	~10^{-4}	~10^{-4}
t_C *(s)*	~$4 \cdot 10^{-7}$	~$2 \cdot 10^{-5}$	~$6 \cdot 10^{-6}$	~$4 \cdot 10^{-4}$
t (s)	~$4 \cdot 10^{-7}$	~$2 \cdot 10^{-5}$	~10^{-4}	~$5 \cdot 10^{-4}$
E_{max} *(J)*	~$4 \cdot 10^{-7}$	~$5 \cdot 10^{-6}$	~$7 \cdot 10^{-6}$	~$8 \cdot 10^{-5}$
P_{max} *(W)*	~1	~0.2	~0.07	~0.16

II. 4. Radioisotope Energy Sources

Electrochemical sources described in previous sections have energy output ~1 eV/atom, which is related to the energy of inter-atomic bonds. In principle, the 'intra-atomic' energy (i.e. of nuclear bonds) is much higher and therefore its utilization seems attractive for embedded micro-power sources in size-constrained systems [6,7]. The energy of radioisotopes is released in energetic particles, typically α- (He ions), β- (electrons) and γ- (electromagnetic radiation) particles. The α- and β- emission could be utilized in energy sources. Several examples of α- and β- radioisotopes are given in Table II.

Table II. Characteristic parameters of several radiosisotopes

Radioisotope	E_{max}, eV	τ (s)	$L(E_{max})$, µm [8]	J/cm^3	W/cm^3
(α) ^{210}Po	$5.4 \cdot 10^6$	$1.7 \cdot 10^7$	~26	$2.3 \cdot 10^{10}$	$1.4 \cdot 10^3$
(α) ^{238}Pu	$5.5 \cdot 10^6$	$4.0 \cdot 10^9$	~27	$4.4 \cdot 10^{10}$	11
(β) ^{3}H	$1.86 \cdot 10^4$	$5.6 \cdot 10^8$	~4	$4.9 \cdot 10^4$	$8.8 \cdot 10^{-5}$
(β) ^{63}Ni	$6.69 \cdot 10^4$	$4.6 \cdot 10^9$	~34	$2.3 \cdot 10^8$	$5.1 \cdot 10^{-2}$

The energy release by radionuclides can be calculated using radioactive decay formula:

$$E(t) = \varepsilon \cdot N(t) = \varepsilon N_0 \exp\left(-\frac{t}{\tau}\right) \qquad (11a)$$

Where N_0 is the initial number of the atoms, $N(t)$ is the number of atoms, which have not released an energetic particle by the time t, τ is the "mean life time" of a radioactive atom and ε is the average energy of the particle released by an radioactive atom. For α-particles emission (discrete energy spectrum), $\varepsilon \approx E_{max}$, while for β-emission (continuous energy spectrum), the average energy of electrons is approximately 1/3 of the maximum energy E_{max}. The total energy released by the radioactive sources is:

$$E = \varepsilon N_0 \qquad (11b)$$

The average power of N_0 radioactive atoms is

$$P = \frac{\varepsilon N_0}{\tau} \qquad (12)$$

It appears that to maximize energy output, the radioisotope with the most energetic particles would need to be used. However, there is a severe constraint on the minimum size of the radioisotope energy sources arising from the ability of matter to absorb radiation. The energy released by radionuclides needs to be captured and converted in a usable form of energy. All existing schemes of energy capturing/conversion are based on the interaction of the energetic particles with absorbing matter. Products of this interaction are: excess charge, electromagnetic radiation and heat.

The absorbing matter is characterized by its "stopping power". The distance L to the point where the energetic particle has lost all its energy is called the range, and it presents a characteristic minimum size of radioisotope energy source. Stopping ranges in Si for α- and β-particles of different energies are given in Table 1. The stopping range can be approximated as:

$$L(\varepsilon) \approx a\varepsilon^n \tag{13}$$

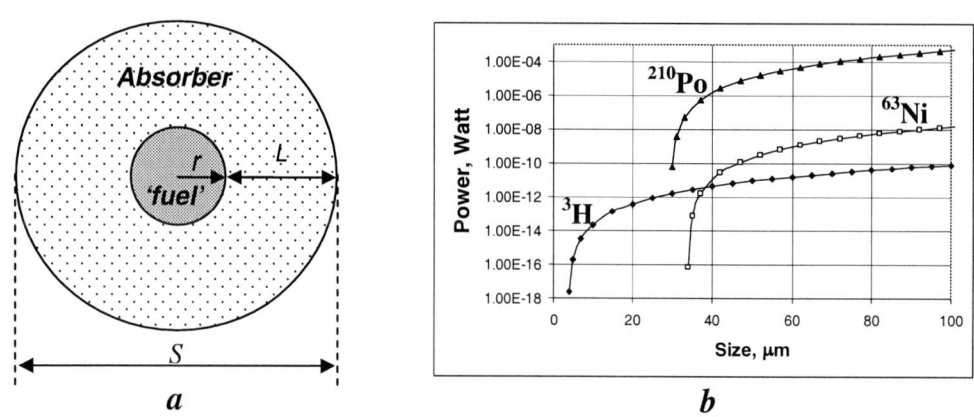

Figure 2. Size constraints of a radioisotope energy source: (**a**) a schematic drawing of the energy source, consisted of a 'fuel compartment' and an 'energy absorber', (**b**) maximum power as a function of total size.

The size of a radioisotope energy source (lower bound) is a sum of the 'fuel compartment' size, r, and the absorber thickness, which is about the stopping range L: $S=r+L$ (Fig. 2a), and the absorber thickness is given by (13). The size of the 'fuel compartment' determines the maximum energy storage, E and power delivery, P of the source:

$$r \sim \left(\frac{E}{\varepsilon n_{at}}\right)^{\frac{1}{3}} = \left(\frac{P\tau}{\varepsilon n_{at}}\right)^{\frac{1}{3}} \tag{14}$$

And thus the lower bound on the size of a radioisotope energy source is:

$$S_{min} = a\varepsilon^n + \left(\frac{P\tau}{\varepsilon n_{at}}\right)^{\frac{1}{3}} \tag{15}$$

Maximum power of radioisotope energy sources as a function of total size is shown in Fig. 2b.

The captured energy must be converted into useful form. In principle, conversion the particle energy into heat allows for ~100% energy collection. However in conversion of the thermal energy into useful work, the upper bound of the useable output energy is given by Carnot efficiency limit:

$$E_{out} = \left(1 - \frac{T_c}{T_h}\right) \cdot E_{ab} = \eta E_{ab} \qquad (16)$$

Assuming T_c=300 K (ambient) and T_h=400 K (a practical operational limit for Si VLSI), the Carnot efficiency η=25%.

Interaction of O- and $\dot{o}$-particles with semiconductor absorbing matter results in creation electron-hole pairs which can be separated at different electrodes, and thus create a potential difference between the electrodes. This is analogous to the effect of incident photons in photovoltaic cells and is referred as to 'alphavoltaiics' and 'betavoltaics'.

One of the factors determining the efficiency bound for betavoltaics is the electron-hole creation energy, $E_{e\text{-}h}$. The *minimum* $E_{e\text{-}h}$ (impact ionization threshold) $E_{e\text{-}h\ min} \sim 3/2 E_g$, resulting from momentum conservation requirement. As was discussed in [9] the *average* energy of electron-hole pair creation can be approximately written as $\overline{E}_{e\text{-}h} \sim 2 E_{e\text{-}h_{min}}$. Next, one needs to take into account that a fraction Δ of the particle kinetic energy goes directly to the lattice vibrations, thus:

$$\overline{E}_{e-h} \sim 3 E_g + \Delta \qquad (17)$$

Relation (17) approximately holds for all semiconductors and for all types of incident particles (e.g. α, β, γ). For example, for silicon, the pair-creation energy is 3.6 eV for α-particles, and 3.8 eV for β–electrons.

The maximum conversion efficiency can be estimated as:

$$\eta = \frac{E_g}{\overline{E}_{e\text{-}h}} = \frac{E_g}{3 E_g + \Delta} = \frac{1}{3 + \dfrac{\Delta}{E_g}} \qquad (18)$$

As follows from (18), higher theoretical efficiency is expected for larger E_g, and the maximum efficiency η_{max}~33% (for silicon, $\ddot{u}$ is~29%).

III. MICRO-SCALE LOGIC UNIT

The autonomous microsystem must have a control unit, which is a specialized micro-scale computer. The capability of the unit will be determined by its complexity (e.g. the device count) and energy of operation.

As the authors have argued in a series of papers [10-12] all binary devices, regardless of the physics of their operation, can be represented by a generic barrier model of Fig. 3. The energy barrier is needed to preserve a binary state in the presence of classic (thermal) and quantum (tunneling) errors (noise). The barrier properties, namely barrier height E_b and barrier width, L, represent the lower bound on the operational energy and size of binary device. The minimum E_b can be estimated from the Boltzmann probability of thermally induced over-barrier transitions:

$$\Pi = \exp\left(-\frac{E_b}{k_B T}\right) \qquad (19a)$$

Requiring P<0.5 (error probability less than 50%), from obtain the minimum E_b:

$$E_b^{min} = k_B T \ln 2 \sim 10^{-21} J \left(T = 300 K\right) \qquad (19b)$$

The minimum barrier width should be sufficient to suppress tunneling, and it can be estimated form the Heisneberg coordinate-momentum relation:

$$\Delta x \Delta p \geq \frac{\hbar}{2} \tag{20}$$

Tunneling is significant when $L \sim \Delta x$. Using $\Delta p \sim \sqrt{2 m_e E_b}$, and (19b), obtain

$$L_{\min} \sim \frac{\hbar}{2\sqrt{2 m_e k_B T \ln 2}} \sim 1 nm \tag{21}$$

The energy E_{bit} to process one bit in practical devices is higher than (19b) due to strict reliability requirement (very low error probability Π, and therefore larger E_b) and a large number of electrons, N_{el} involved in each switching event:

$$E_{bit} \sim N_{el} \cdot E_b \tag{22}$$

There are three main factors that determine E_b and N_{el}, which are outlined below.

System reliability costs: Requirement that all N devices in the logic system operate correctly raises E_b much higher than (19b). In fact E_b is a function of specified error probability Π and the number N of devices in the unit: $E_b = f(\Pi, N)$ [11, 19].

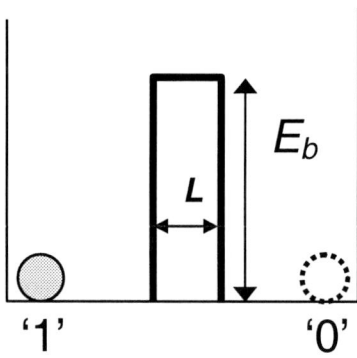

Figure 3. A barrier model for an arbitrary binary switch

Fan-Out costs: The need of each device to communicate to several others (typically four) requires increased length of the short interconnect, and therefore larger number of electrons needed for reliable communication. It was estimated that for 2D system topology and for fan-out of four, a minimum $N_{el}=14$ for 50% probability of correct communication and $N_{el}=42$ for 99% reliability [12, 19]. One the other hand, for a hypothetical 3D system topology implemented with 1D device components, the fan-out costs could be reduced by approximately factor of 10 [11,19].

Long communication costs. Communication at distance is a very costly process. Scaling system size from e.g. 1 cm to 10 μm would dramatically reduce these costs [19].

A summary of projected characteristic device size (L), operation energy (E_{bit}) , device density(n) and the total number of devices in system are given in Table III.

Based on the ITSR projections [13], if the control unit of the nanomorphic cell is implemented by using end-or-of-the-roadmap CMOS technology, it could contain 10^4 transistors (Table III). As a reference, Intel 8080 microprocessor contained only 4500 transistors. Therefore a 10-μm processor could have remarkable complexity. For a limited energy supply of $\sim 10^{-5}$ J as estimated in previous section, the total number of binary events in the control unit is:

Table III. Projected device characteristics for binary switches for different system size and topology

	System size	**System Topology**	E_{bit}	**L, nm**	**N, cm^{-2}**	**N$_{tr}$**	**Ref**
Lower bound		2D, 3D	$\sim k_B T = 3 \cdot 10^{-21}$ J	1	$\sim 10^{13}$	-	[10-12]
2022 CMOS	$\sim$cm	2D	$600 k_B T \sim 2 \cdot 10^{-18}$ J	5	10^{10}	10^{10}	[13]
2022 CMOS	$\sim$10 μm	2D	$260 k_B T \sim 8 \cdot 10^{-19}$J	5	10^{10}	10^4	[19]
2022 CMOS	$\sim$10 μm	3D	$35 k_B T \sim 10^{-19}$ J	5	10^{10}	10^4	[19]

$$N_{bit} = \frac{E_{stored}}{E_{bit}} \sim \frac{10^{-5} J}{10^{-18} J/bit} \sim 10^{13} \text{ binary operations} \qquad (23)$$

These estimates suggest that the micro-scale processor could have remarkable complexity and perform extensive information processing. Note the potential of topology optimization for energy reduction: Quasi 1D (e.g. nanowire) components arranged in 3D structures could considerably reduce the energy due to smaller fan-out costs. Also, embedded non-volatile memory will almost certainly be required and scalability of existing non-volatile memory technologies is an open question at the present time. This may offer the opportunity for application of emerging memory technologies such as nanowire or molecular memories in the implementation of the control unit.

In conclusion, aggressive CMOS scaling is mandatory for implementation of micron-scale systems. Beyond-CMOS devices also need to be analyzed for their potential to offer more functionality at less device count.

IV. NANOMORPHIC CELL COMMUNICATION UNIT

Communication is an essential component of autonoumous microsystems. Here we consider a case of uniformly radiated electromagnetic communication between a 10 μm-nanomorphic cell and an external device located at the distance, r, from the cell (Fig. 4). In the limits, at least one photon must absorbed by the external detector. If the location of the external device relative to the cell is unknown, in order to guarantee at least one photon reach the detector, the entire sphere of radius r must be 'covered' with photons. Area 'covered' by one photon of wavelength λ is $\sim \lambda^2$, and the total number of photon needed is:

$$N_{ph} \sim \frac{4\pi r^2}{\lambda^2} \qquad (24)$$

The energy of each photon is

$$E_{ph} = h\nu = \frac{hc}{\lambda} \qquad (25)$$

Thus, the energy associate with one full 'communication packet', i.e. the minimum energy required to transmit one bit of information such that it is equally accessible at all points on the sphere is

$$E_{com} = N_{ph} \cdot E_{ph} \sim \frac{4\pi r^2}{\lambda^2} \cdot \frac{hc}{\lambda} = \frac{4\pi hc r^2}{\lambda^3} \qquad (26)$$

The size of transducer (e.g. antenna) must be about radiated wavelength λ. Since the transducer size is limited by the cell size (10 μm in our case). For λ=10 Ãm and the distance between the cell and the receiver r=1 m (26) gives E_{com}~2.5 10^{-9} J/bit. (Note, that this energy estimate is a lower bound on communication. It doesn't consider, e.g. efficiencies of both tranducer and detector, noise etc.). This result reveals that communication by the extreme microsystem might be very costly. For example for the total available energy ~10^{-5} J (see section II) and E_{com}~2.5 10^{-9} J/bit, the cell could send maximum 4000 bits.

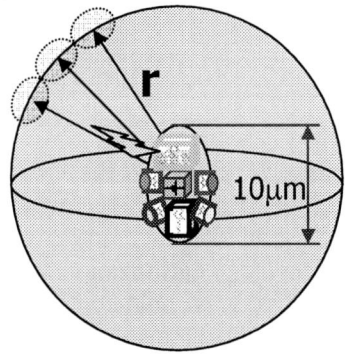

Figure 4. Illustration to communication between a nanomorphic cell and an external device.

Several options could be considered to minimize the energy expeditures for communication. First as follows from (26) E_{com}~$1/\lambda^3$ and therefore longer antenna would dramatically reduce communication energy. Thus external 'tail' antennas is an attractive option. Carbon nanotubes might be the best material for such antennae [20] as they offer suitable length range, smaller diameter, and good conductive properties.

Directional transmission would reduce the number of photons in the packet and therefore the total energy. Of course, the orientation problem must be solved. One not very elegant solution would be to use as many external detectors as possible.

Finally, comparison of the costs to 'process' (10^{-18} J/bit) and 'communicate' (10^{-9} J/bit) of information suggests that overall design goal should be to the communication and to maximize the 'intelligence' of the nanomorphic cell.

V. BIO-SENSORS AT THE MICRO-METER SCALE

An essential function of the extreme microsystem for bio-applications is to transform biological signals into electrical form for subsequent processing and analysis to provide a basis for further actions. Many sensors can be regarded as binary switches (Fig. 3) , whose barrier is deformed by different stimuli other than charge, e.g. *mechanical, optical, thermal, chemical*. In principle, the sensor can be powered, at least partially, by the energy of the external stimulus. For the purposes of this study, the sensors must be of size commensurate with that of the artificial cell. Typically, bio-sensors are comprised of an element that responds to biological signals and an element that generates electrical signals in response to the biological signals. In the following, we offer a few examples of promising current research for bio-sensors at the micro-meter scale.

IV.2. Electrical sensors

Electrical interfaces between nerve cells and semiconductor microstructures is a very important topic for e.g. prosthetic devices based on recording and stimulating neuroelectric activity. The main technological issue of electrical biosensors is that the charge carriers in the living cells are ions and in solid state devices, the charge carriers are electrons. Therefore integration between microionics and microelectronics is needed [14-16].

The electrical elements of living cells are voltage-gated ion channels. The electrical activity of a cell is accompanied by the opening of the channels in the cell's membrane which allows ions to flow to the cell's exterior. The time scale of this 'ionic' event is ~ms [14-16]. The current flowing through the ion channels causes a voltage change in the electrolyte, which can be sensed by gate-less FETs. In addition, if an ion sensitive material such as silicon dioxide is used as a gate isolator, the FET will sense the change of ion concentration.

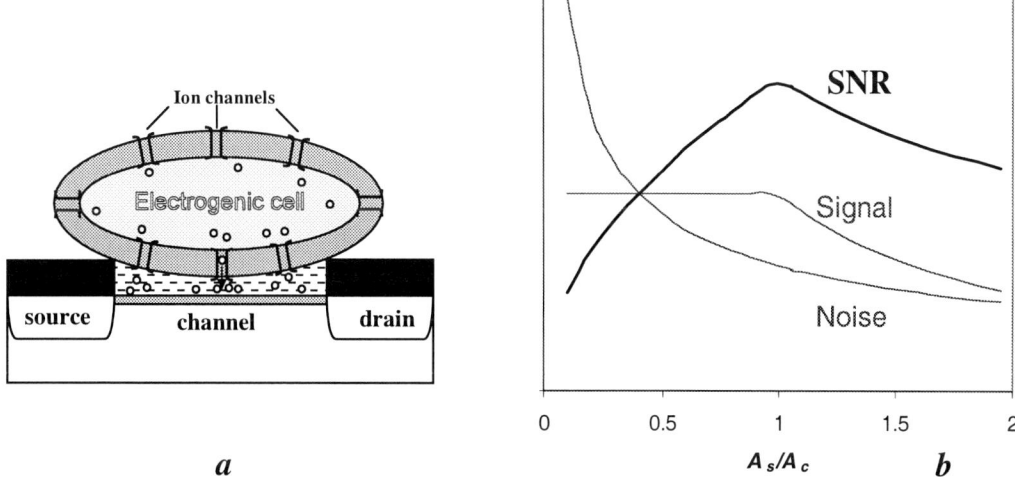

Figure 5. Recording of electrical activity of a cell with FET biosensor: **(a)** schematic of FET/cell configuration, **(b)** - Scaling behavior of signal-to-noise ratio of FET biosensor

To sense extracellular electrical activity, the cell must be brought in proximity to the FET channel, and the extracellular electrolyte plays the role of an FET gate (Fig. 5a). Note that there is a gap or cleft in the junction between the cell membrane and FET sensing surface. The cleft is filled with electrolyte; a typical separation is ~50-70 nm [14-16].

FET biosensors have been successfully used to record electrical signals from individual cells [14-16]. The scaling-sensitivity trade-offs of FET biosensors has been analyzed [14-16]. Based on signal-to-noise-ratio (SNR) considerations, it was argued that the optimum 'sensing' area A_s of the gate/channel region should be close to the cell size. The arguments can be summarized in a simplified form as follows:

There are two important size-dependent sources of noise:
1. The Ohmic resistance of electrolyte between the attached cell membrane and the transistor that contributes to the thermal noise: $P_n \sim R \sim 1/A_s$ ($V_n \sim \sqrt{P_n} \sim 1/\sqrt{A_s}$)

2. Ion channels in the cell open and close (quasi-) randomly, generating fluctuations in the ion concentration in the cleft (shot noise-like process). The larger the number, N, of channels in the contact area, the lower the fluctuations: $P_n \sim 1/N \sim 1/A_s$ ($V_n \sim \sqrt{P_n} \sim 1/\sqrt{A_s}$)

On the other hand, the signal voltage is proportional to the gate charge density, $V \sim q/A_s$. The gate charge q is proportional to the number of ionic channels, N, in the 'sensing' area, which in turn is directly proportional to the gate area A_s. If A_s is less than the cell size A_c, $V \sim A_s$. When $A_s > A_c$, the gate charge q remains constant, and $V \sim q/A_s$ decreases. As a result, the signal-to-noise ratio has a maximum at $A_s = A_c$, as is illustrated in Fig. 5b.

Examples of typical gate dimensions of practical FET biosensors are 2 μm × 20 μm, 6 μm × 7 μm and 22 μm × 24 μm [15,16]. The RMS noise in the range 2 Hz-2kHz is ~5 μV for larger transistors and 14 μV for smaller devices [15]. Larger FET allows for "full-cell" recording, while smaller devices can be used to record e.g. vesicle releases [16]. It is projected that for detection of small vesicles with diameter ~50 nm, a FET sensor with gate area ~0.5 μm^2 will be needed. In this case, the projected signal amplitude is ~400μV at a detection limit ~300 μV. The latter example could be regarded as an approximate practical scaling limit for FET biosensors.

IV.3. Chemical and Bio-chemical sensors

An ion-sensitive FET can also be used to analyze the chemical composition of a solution. The simplest goal would be to measure pH level, which is related to the concentration of hydrogen ions. The ionic activity of the cell can be measured by the ion-sensitive FET (ISFET) described in the previous section. Typical ISFET detection limit is ~μM, while aM concentrations are characteristic for individual cell activity. Sensor sensitivity could be dramatically enhanced by using a nanowire (NW) channel FET, due to very high specific surface and small cross-sectional area of the nanowire. In a conventional (2D) FET, the gate charge changes the channel conductance only in a thin interface region, while in NW devices, the conductance is modulated at the "bulk" level, resulting in higher sensitivity [14]. It can be shown that the ion sensitivity of NW FET increases as the NW radius decreases [17, 18]. Below a simple analysis to illustrate this point is offered for p-type doped NW FET pH sensor.

The current through nanowire is proportional to the cross-sectional area of the NW channel:

$$I_{NW,0} = \frac{V}{R} = \frac{VA}{\rho L} = \frac{V \cdot \pi r^2}{\rho L} \tag{27}$$

When H$^+$ ions are adsorbed on the NW surface, the NW mobile charge carriers become depleted in the surface region of width W, which reduces the conductive cross-section of the NW as shown in Fig. 6. The corresponding current through NW will be:

$$I_{NW} = \frac{V \cdot \pi (r-W)^2}{\rho L} \tag{28}$$

The relative change of current due to adsorption of ions is a measure of device sensitivity and is given as (assuming $W \ll r$):

$$\frac{\Delta I_{NW}}{I_{NW,0}} = \frac{r^2 - (r-W)^2}{r^2} = \frac{2rW - W^2}{r^2} \approx \frac{2rW}{r^2} = \frac{2W}{r} \tag{29}$$

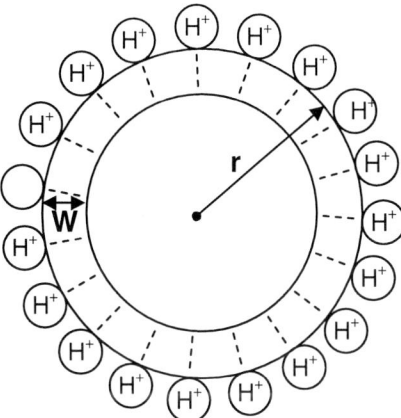

Figure 6. A schematic cross-section of NW showing external surface charge and internal charged depletion layer

The depletion width can be estimated based on the charge neutrality condition, i.e. the total positive charge of ions adsorbed at the surface is equal to the total negative charge of ionized acceptors in the depleted volume of NW:

$$q_s = q_d$$
$$n_s \cdot 2\pi r L = N_a \cdot 2\pi r L W$$

(30)

From (30) the depletion width W can be found as:

$$W = \frac{n_s}{N_a}$$

(31)

Putting (31) into (29) we obtain the NW sensor sensitivity:

$$\frac{\Delta I_{NW}}{I_{NW,0}} \approx \frac{2}{r} \frac{n_s}{N_a}$$

(32)

Because of their high sensitivity, the nanowire FET shows promise for both chemical and biochemical microsensors. Si-NW FET's for pH sensing have been demonstrated in [17, 18]. Typical NW sensor dimensions are ~50-100 nm width and several μm in length [17,18]. In [17] the NW pH sensor was tested in the pH range 2-12, and sensitivity of ~80 nS/pH was achieved. Another application for these devices is the direct detection of macromolecules. This application requires that the nanowire be 'functionalized' by binding e.g., a receptor protein or a single-stranded DNA to the surface of NW sensor [18]. Such a 'functionalized' surface will preferentially adsorb only certain types of molecules. In [18] a functionalized NW sensed proteins at the 10 fM levels at SNR of 140. The projected detection floor is ~70 aM. Application of NW FET biosensors for immunodetection using antibodies has been studied. The ability to detect antibodies at less than 100 fM has been demonstrated [18].

CONCLUSIONS

We have offered a feasibility-level examination of the physical limits for various subsystems for a nanomorphic cell that are subjected to extreme volumetric scaling. Our results have been surprisingly affirming with respect to the possibility of of such a system providing useful functions. One observation is that 1D structures could play an important role in the realization of such a system since they appear to offer unique advantages in each of the subsytsems. A significant challenge is that of actually fabricating the nanomorphic cell since planar fabrication technologies used in the semiconductor industry might not suffice.

ACKNOWLEDGMENTS

The authors wish to acknowledge the stimulating discussions and inputs provided by the participants at the NSF/SRC Forum on Nanomorphic Systems in November, 2007.

REFERENCES

1. A. Heller, *Anal. Bioanal. Chem.* **385,** 469 (2006)
2. A. Heller, *Phys. Chem. Chem. Phys.* **209,** (2004)
3. D. Wu, R. Tucker, and H. Hess, *IEEE Trans. Adv. Pack.* **28,** 594 (2005)
4. R. Kötz and M Carlen, *Electrochem. Acta* **45,** 2483 (2000)
5. A.Lewandowski and M. Galisnki, *J. Power Syst.* **173,** 822 (2007)
6. M. V. S. Chandrashekhar, R. Duggirala, M. G. Spencer, and A. Lal, *Appl. Phys. Lett.* **91,** 053511 (2007)
7.. C. J. Eiting, V. Krishnamoorthy, S. Rodgers, T. George, J. D. Robertson, and J. Brockman, *Appl. Phys. Lett.* **88,** 064101 (2006)
8. *Stopping-Power and Range Tables for Electrons, Protons, and Helium Ions,* http://physics.nist.gov/PhysRefData/Star/Text/contents.html
9. C. A. Klein, *J. Appl. Phys.* **39,** 2029 (1968)
10. V. V. Zhirnov, R. K. Cavin, J. A. Hutchby, and G. I. Bourianoff, *Proc. IEEE* **91,** 1934 (2003)
11. R. K. Cavin and V. V. Zhirnov, "*Keynote*: Nano-Architecture Challenges: A Fundamental Physics Perspective", *IEEE/ACM Intern. Symp. on Nanoscale Architectures, San Jose, CA, October 21-22, 2007*
12. R. K. Cavin and V. V. Zhirnov, *Solid-State Electron.* **50,** 520 (2006)
13. *The International Technology Roadmap for Semiconductors,* 2007; http://www.itrs.net/
14. M. Voelker and P. Fromherz, *Small* **1,** 206 (2005)
15. M. Voelker and P. Fromherz, *Phys. Rev. Lett.* **96,** 2281102 (2006)
16. J. Lichtenberger and P. Fromherz, *Biophysical J.* **92,** 2262 (2007)
17. I. Park, Z. Y. Li, X. Li, A. P. Pisano, and R. S. Williams, *Biosensors and Bioelectron.* **22,** 2065 (2007)
18. Eric Stern, J. F. Klemic, D. A. Routenberg, P. N. Wyrembak, D. B. Turner-Evans, A. D. Hamilton, D. A. LaVan, T. M. Fahmy, and M. A. Reed, *Nature* **445,** 519 (2007)
19. V. V. Zhirnov, "'Minimizing' the effects of Physical Limits", SRC/NSF Forum on Nano-Morphic Systems: Processes, Devices, and Architectures, November 8-9, 2007, Stanford University, Stanford, CA.
20. Q. Zhu, L. Wu, S. Sheng, Z. C. Mei, W. F. Liu, W. L. Cai, and L. Z. Yao, J. Vac. Sci. Techol. B **25,** 1630 (2007)

Mater. Res. Soc. Symp. Proc. Vol. 1067 © 2008 Materials Research Society

1067-B01-04

Logic Devices with Spin Wave Buses - an Approach to Scalable Magneto-Electric Circuitry

Alexander Khitun[1], Mingqiang Bao[1], Yina Wu[1], Ji-Young Kim[1], Augustin Hong[1], Ajey P Jacob[2], Kosmas Galatsis[1], and Kang L Wang[1]

[1]Electrical Engineering, University of California Los Angeles, Los Angeles, CA, 90095-1594

[2]TMG External Programs, Intel Corporation and Western Institute of Nanoelectronics, Los Angeles, CA, 90095-1594

ABSTRACT

We analyze spin wave-based logic circuits as a possible route to building reconfigurable magnetic circuits compatible with conventional electron-based devices. A distinctive feature of the spin wave logic circuits is that a bit of information is encoded into the phase of the spin wave. It makes possible to transmit information as a magnetization signal through magnetic waveguides without the use of an electric current. By exploiting sin wave superposition, a set of logic gates such as AND, OR, and Majority gate can be realized in one circuit. We present experimental data illustrating the performance of a three-terminal micrometer scale spin wave-based logic device fabricated on a silicon platform. The device operates in the GHz frequency range and at room temperature. The output power modulation is achieved via the control of the relative phases of two input spin wave signals. The obtained data shows the possibility of using spin waves for achieving logic functionality. The scalability of the spin wave-based logic devices is defined by the wavelength of the spin wave, which depends on the magnetic material and waveguide geometry. Potentially, a multifunctional spin wave logic gate can be scaled down to $0.1 \mu m^2$. Another potential advantage of the spin wave-based logic circuitry is the ability to implement logic gates with fewer elements as compared to CMOS-based circuits in achieving same functionality. The shortcomings and disadvantages of the spin wave-based devices are also discussed.

INTRODUCTION

The rapid approach to the scaling limit of metal-oxide semiconductor field-effect transistor (MOSFET) has stimulated a great deal of interest to research alternative technologies, which may overcome the constrains inherent to CMOS-based circuitry and provide a route to more scalable and less power consuming logic devices. One of the most promising approaches is in the use of spin as a state variable. There is an impetus for the development of novel spin-based logic circuits that are aimed to provide high information/signal processing rates for lower power dissipation and scaleable to the nanometer range. Dipole-dipole and exchange interaction can be used for information transmission among spin-based devices without the use of an electric current. There are several approaches to spin-based devices such as Magnetic Cellular Automata [1], Domain-Wall Logic [2], and more recently devices with Spin Wave Bus [3]. In our preceding works [3-5], we have developed the general concept of logic circuits with Spin Wave Bus. Briefly, the basic idea to use magnetic films as spin conduit of wave propagation or referred to as – Spin Wave Bus, where the information can be coded into a phase of the propagating spin wave. A first working spin-wave based logic circuit has been experimentally demonstrated by M. Kostylev et al. [6]. Recently, the functionality of spin-wave logic exclusive-not-OR and not-AND gates based on a Mach-Zehnder-type spin-wave interferometer has been realized [7]. In this work, we describe our original design of the three-terminal spin wave circuit and present experimental data illustrating its performance. The main advantage of the presented design is that the length of the circuit is no longer limited by the phase accumulation length [6, 7]. The latter translates in the intrigue possibility to scale down the length of the whole circuit to the wavelength of the spin wave (10nm-100nm). We analyze the potentials and constrains of the spin wave-based devices for scalable "beyond CMOS" logic circuits.

PRINCIPLE OF OPERATION

The operation of the spin wave-based logic circuit includes three major steps: (i) conversion of the input data into spin wave signals, (ii) computation using spin waves, and (iii) conversion the result of computation in the form of output voltage signals. The distinctive feature of the spin wave-based logic circuit is that a bit of information is encoded into the *phase* of the spin wave.

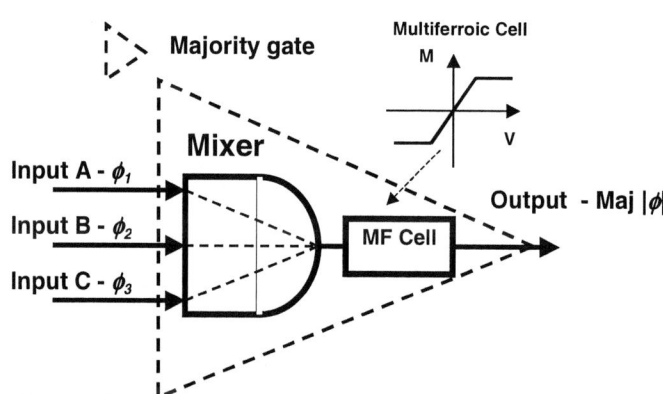

Fig.1 Majority logic gate based on spin waves interference. Three spin waves having initial phases 0 or π are mixed and amplified by a ME cell. The phase of the output signal is a majority of the input phases.

Two relative phases of "0" and "π" may be used to represent two logic states 1 and 0. Being excited, spin waves propagate throughout the structure consists of spin waveguides. The data processing in the circuit is accomplished by manipulating the relative phases of the propagating spin waves. Finally, the result of computation is converted in the voltage signal.

The utilization of wave interference provides an effective way to achieve logic functionality. The interference of several spin waves results in the magnetization change, which is the sum of the separate amplitudes of each wave. Similar to other waves, spin waves can interfere in a constructive (in phase) or a destructive (out of phase) manner. The amplitude of the resultant signal depends on how many waves are coming in phase or out of phase. In order to equalize the output amplitude, a nonlinear element (for example, a magneto-electric cell) should be added to the circuit. In figure 1 we have schematically shown a three-input one-output logic device, which consists of three waveguides combined with the nonlinear element. Three input spin wave signals A, B, and C (phases ϕ_1, ϕ_2 and, ϕ_3 respectively) of the same frequency are mixed, and the output signal is amplified and equalized by the magnetoelectric cell (ME Cell). The phase of the output spin wave always corresponds to the majority of the three input signals, thus, the device operates as a Majority logic gate. The same device structure can be used as a two-terminal AND or OR logic gate by using the third input C as a control. For example, if the logic state of the control input C is 0, the circuit operates as the AND gate. If the logic state of the control input C is 1, the circuit operates as the OR gate. In order to illustrate the utilization

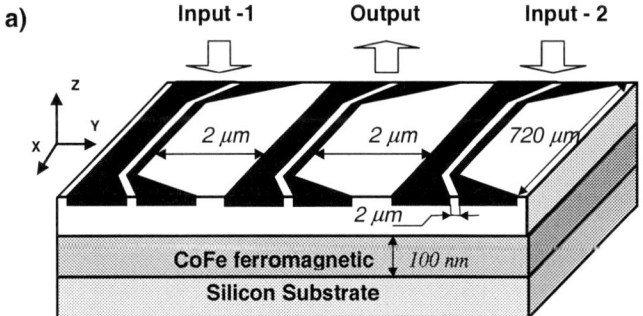

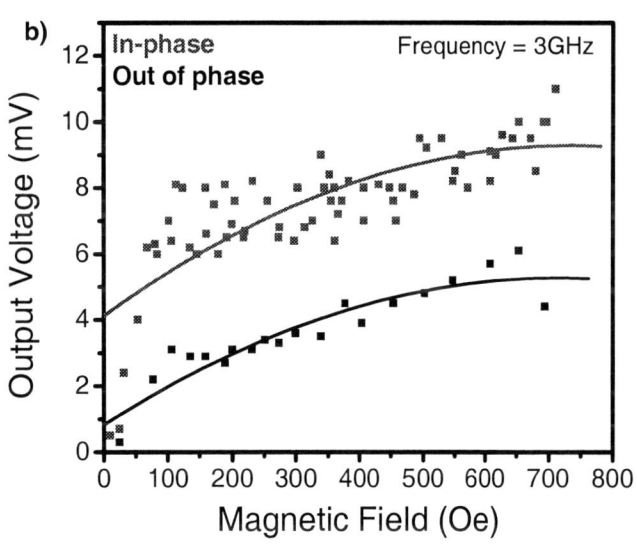

Fig.2(a) The general view of the three-terminal spin wave device. The core of the structure from the bottom to the top consists of a silicon substrate, a 100nm thick CoFe film, and a 300nm thick silicon dioxide layer. Three ACPS lines on the top are used as the two input (edge) and one output (in the middle) ports. (b) Experimental data. Output signal amplitude for two spin waves coming in-phase (red curve) and out of phase (black curve).

of spin wave superposition, we present experimental data on a prototype spin wave device.

EXPERIMENTAL DATA

In figure 2 (a) we show the general view of the prototype spin wave-based logic device. The core of the structure from the bottom to the top consists of a silicon substrate, a $Co_{30}Fe_{70}$ 100nm thick film, and a 300nm thick silicon dioxide layer. The film was deposited using a high vacuum rf-sputtering system and the film exhibited a saturation magnetization (B_s) of ~2.2 T. There are three asymmetric coplanar strip (ACPS) transmission lines on the top of the structure. The edge ACPS lines are the transducers to excite spin waves, and the line in the center is the receiver to detect the inductive voltage produced by two spin wave signals. The distance between the microstrips is 4μm. The ACPS lines are connected to the VNA working at frequency from 500MHz to 8500GHz, and a uniform static magnetic field up to 940 Oe is applied in the plane and perpendicular to the wave vector (MSSW configuration). In order to exclude direct electro-magnetic coupling between the transducer and the receiver lines, all experiments had been repeated with and without magnetic field, and the spin wave signals are obtained by the subtraction. All measurements are carried out at room temperature.

In figure 2(b), we present experimental data showing output inductive voltage measured at the central ACPS line at different values of the external magnetic field (excitation frequency 3GHz). The red and black curves depict the output power for the in-phase ($\Delta\phi=0$) and out of phase ($\Delta\phi=\pi$) cases, respectively. The initial phase difference between two input signals is defined by the direction of the current flow in the excitation lines. The phase difference is zero if the direction of current flow is the same (clockwise or counter-clockwise wise) in both lines. If the directions in the excitation lines are such as one loop is clockwise and the other is counter-clockwise, the spin wave signals receive a π relative phase difference. These experimental data show a prominent (about 4 times) output power difference. We would like to stress that the observed output power modulation is achieved by the control of the relative *phases* of two input signals.

DISCUSSION

The scalability of the spin wave-based logic devices is defined by the wavelength of the spin wave. The length of a logic gate has to be an integer number of the wavelength. It restricts the minimum length of a waveguide per circuit to be at least one wavelength long. In contrast, the width and the thickness of the spin waveguides can be scaled down to several nanometers. The size of the magnetoelectric cell (the area covered by piezoelectric) can be as small as a one magnetic domain. The permissible size variation of the magnetoelectric cell and spin wave splitters/combiners, which does not effect logic functionality, can be estimated as $\lambda/8$. Any defect of the waveguide structure with characteristic size much smaller than the wavelength has no critical effect on spin wave propagation. To scale down the length of the logic gate one needs to decrease the wavelength. At the same time, the shorter wavelength signal becomes more sensitive for structure imperfections. The optimum wavelength has to be found taking into consideration particular material structure.

The power consumption of the spin-wave based logic circuit is defined by the energy losses during spin wave excitation, propagation, and amplification. The energy is dissipated in the conducting wires used for magnetic field generation, in the spin-wave bus during the spin wave propagation, and in the magnetoelectric cells. The energy of the spin wave signal can be any close to kT and limited by the thermal noise only. For an acceptable error rate, it can be estimated to be about $20kT$ per bit. However, the energy required for spin wave excitation and amplification may be several orders of magnitude higher depending on the spin wave excitation mechanism and the strength of the magnetoelectric coupling of the ME cell. Detailed numerical assessments on power dissipation in ferromagnetic spin wave bus during signal propagation are given in [8]. Based on this model, we estimate the energy per function in the spin-wave logic circuit to be $10^5 kT$. The lack of experimental data on the spin wave amplification via the magnetoelectric coupling does not allow us to conclude on the practically achievable minimum.

The demonstrated prototype logic device is a step toward to spin wave-based logic circuitry. The utilization of wave properties such as superposition and interference opens a new horizon for logic devices with potential capabilities far beyond current CMOS-based circuits. In general, Majority logic is more powerful for implementing a given digital function with a smaller number of logic gates [9]. For example, the full adder may be constructed on three majority gates and three inverters In contrast, a Boolean-based implementation requires a larger circuit with seven or eight gate elements (about 25–30 MOSFETs) [10].

Spin wave based devices posses two important disadvantages: (i) low signal propagation speed (10^4-10^5m/s), and (ii) high spin wave attenuation (nanosecond scale decay time [11]). All together, these disadvantages limit possible spin wave utilization only for intra-chip communication purposes. As a physical mechanism for information transition, spin waves can, potentially, compete with the conventional charge-transfer approach at the ultra-high chip density – more than 10^{10} devices per square centimeter.

CONCLUSIONS

The use of spin waves offers an original way to build a new type of reconfigurable spin-based logic devices. We have demonstrated a room temperature working three-terminal prototype logic circuit utilizing spin waves for information transmission and processing. The output power modulation is achieved via the control of the relative phases of two spin wave signals. The obtained data shows the possibility of using spin waves for achieving logic functionality and for building scalable circuits integrated on a silicon platform. The issues of low group velocity and damping restrict potential spin wave applications for micrometer short range.

ACKNOWLEDGMENTS

We would like to thank Dr. S. Wang and D.W. Lee (Stanford University) for CoFe deposition. The work was supported in part by the Focus Center Research Program

(FCRP director: Dr. Betsy Weitzman) Center of Functional Engineered Nano Architectonics (FENA), and by the Nanoelectronics Research Initiative (NRI Director: Dr. Jeff Welser) - The Western Institute of Nanoelectronics (WIN).

REFERENCES

[1] S. Bandyopadhyay and V. P. Roychowdhury, "Granular nanoelectronics," *IEEE Potentials*, vol. 15, pp. 8-11, 1996.

[2] D. A. Allwood, G. Xiong, C. C. Faullkner, D. Atkinson, D. Petit, and R. P. Cowburn, "Magnetic domain-wall logic," *Science*, vol. 309, pp. 1688-92, 2005.

[3] A. Khitun and K. Wang, "Nano scale computational architectures with Spin Wave Bus," *Superlattices & Microstructures*, vol. 38, pp. 184-200, 2005.

[4] K. L. Wang, A. Khitun, and A. H. Flood, "Interconnects for nanoelectronics," presented at Proceedings of the IEEE 2005 International Interconnect Technology Conference (IEEE Cat. No. 05TH8780). IEEE. 2005, pp. 231-3. Piscataway, NJ, USA.

[5] A. Khitun and K. L. Wang, "Nano logic circuits with spin wave bus," presented at Proceedings. Third International Conference on Information Technology: New Generation. IEEE Computer Society. 2006, pp. 6. Los Alamitos, CA, USA.

[6] M. P. Kostylev, A. A. Serga, T. Schneider, B. Leven, and B. Hillebrands, "Spin-wave logical gates," *Applied Physics Letters*, vol. 87, pp. 153501-1-3, 2005.

[7] T. Schneider, A. A. Serga, B. Leven, B. Hillebrands, R. L. Stamps, and M. P. Kostylev, "Realization of spin-wave logic gates," *Appl. Phys. Lett.*, vol. 92, pp. 022505-3, 2008.

[8] A. Khitun, D. E. Nikonov, M. Bao, K. Galatsis, and K. L. Wang, "Efficiency of spin-wave bus for information transmission," *IEEE Transactions on Electron Devices*, vol. 54, pp. 3418-21, 2007.

[9] A. R. Meo, "Majority Gate Networks," *IEEE Transactions on Electronic Computers*, vol. EC-15, pp. 606-18, 1966.

[10] T. Oya, T. Asai, T. Fukui, and Y. Amemiya, "A majority-logic device using an irreversible single-electron box," *IEEE Transactions on Nanotechnology*, vol. 2, pp. 15-22, 2003.

[11] M. Covington, T. M. Crawford, and G. J. Parker, "Time-resolved measurement of propagating spin waves in ferromagnetic thin films," *Physical Review Letters*, vol. 89, pp. 237202-1-4, 2002.

AUTHOR INDEX

Bao, M. 33

Biesemans, S. 12

Bouvet, D. 7

Caro, J. 12

Cavin, R. 18

Collaert, N. 12

Galatsis, K. 33

Gysel, R. 7

Hollenberg, L. 12

Hong, A. 33

Ionescu, M.A. 7

Ishibashi, A. 1

Jacob, A.P. 33

Kaiju, H. 1

Khitun, A. 33

Kim, J. 33

Klimeck, G. 12

Kondo, K. 1

Lansbergen, G. 12

Rahman, R. 12

Riester, S. 7

Rogge, S. 12

Salvatore, G.A. 7

Setter, N. 7

Stolichnov, I. 7

Wang, K.L. 33

Wellard, C. 12

Wu, Y. 33

Zhirnov, V. 18

Cambridge University Press
32 Avenue of the Americas, New York, NY 10013-2473, USA

Materials Research Society
506 Keystone Drive, Warrendale, PA 15086

ISBN 978-1-60560-845-7